An Introduction to Environmental Science

Dr. R.M. Mishra

SHREE PUBLISHERS & DISTRIBUTORS
NEW DELHI-110 002

Edition : 2009

Published by :

SHREE PUBLISHERS & DISTRIBUTORS

22/4735 Prakash Deep Building,

Ansari Road, Darya Ganj,

New Delhi–110 002

ISBN : 978-81–8329–332–7

Printed by :

Mehra Offset

Delhi–110 006

PREFACE

The study of *Environmental Science* is a new concept but its basic nature is as old as human civilization. Environment is the aggregate of external conditions that influence the life of an individual or the population specifically the life of man. Ultimately, *Environment* determines the quality of survival of man on earth. Man is supposed to improve his *Environment* because there is environmental pollution around him. It also effects all sorts of creatures and species.

Environment is a broad concept to surround the whole range of diverse surroundings in which we perceive, experience and react to events and changes. Basic components of *Environment* include Earth, Land, Water, Forest, Air and Extraction of Minerals etc. which need not be polluted, wasted and exhausted.

The present book elaborately discusses the various components of the Environment and Zones of Atmosphere, major components of Eco-systems and its structure as well as Organisms represented in food chains.

Natural Energy Resources and its various types are vital to the survival and development of human population. Natural Resources are of two types Renewable and Non-Renewable. Solar energy Wind energy, Water energy, Biomass are Renewable, whereas Natural Gas, Coal, Oil & Petroleum products etc. are non-Renewable. Balanced use and extractability of the material available and the demand for them must be taken into consideration for human existence in view of the great threat to the *Environment*.

The book also discusses the Diversification of Biotic environment. The Biodiversity found on Earth consists of many millions of distinct biological species. Thus, the *Environmental Pollution* problems have become a global issue. This books also focuses on major categories of Environmental Pollution—Air, Water, Soil, Marine, Noise, Theism, Nuclear and Solid Waste etc. Environmental Pollution prevention is to protect our health and the environment in which we live in. Protective measures no doubt, protect human race and all sorts of species on earth.

In the concluding part, the writer has emphasized the need of *Environmental Awareness in Society* besides various Environmental Protection Acts. We hope that the points discussed in this book will help the readers to understand various aspects of *Environmental Science* in a most comprehensive and critical way to face the present and future challenges in this field.

—Dr. R.M. Mishra

1

ENVIRONMENT

Environment is taken to mean all those which are physical,chemical,organic and physical, chemical,organic and non-organic components of the atmosphere,lithosphere and oceans. Environment is the aggregate of external conditions that influence the life of an individual or the population specifically the life of man;environment ultimately determines the quality of survival of life.

Environment consists of various types of forces like physical,intellectual,social,moral,economic, political and cultural; forces.Environment is the aggregate of all the external forces,influences and conditions that affects the life,nature,behaviour and the growth,development and maturation of living organisms.

The Environment consists of natural as well as socio-culture environment.Man has to improve the quality of his environment because there is environmental pollution or crisis.It is due to over consumption of natural resources over population,industrialisation and unscientific attitude of human beings.

Multidisciplinary Nature of Environmental Studies

Environment refers the surroundings of an organisms which

includes both living and non living components.The word meaning of environment is to surround and to develop.

Environmental study is a new concept but its basic nature is old as human civilisation since early days.The basic ingredients of natural environment and its pollutants which should be included in the study of environmental education at various levels of our education system.

These are the following disciplinary nature of environmental study:

1. **Education nature** is the first disciplinary nature of environmental studies.

Education is an independent field of study or discipline which is concerned with process of development,training and instruction. The job of teacher is to create a conductive evironment in the school and in the classroom to have the desirable change in the behaviour of child.

A child gains experience from the school environment(physical and social)and from the classroom climate(social and emotional). Experience means learning or modification of behaviour. The experience can only be provided through environment. In the classroom a teacher performs some activities or teaching to generate social and emotional climate with the help of some content for providing new experiences to his students. The educational institutions are located outside the locality to have good physical and biological environment of students.

The Environment cannot induce the traits and abilities which are inborn of an individual.Environment can develop inborn traits and abilties. Environmental education includes both fornal and informal education.

In formal education four different inter-related components are recognised:-

1. Awareness of environment
2. Environment and real life situation
3. Conservation of resources
4. Sustainable development of individual

These components are related to the four levels of education--

1. Primary Education-awareness
2. Secondary Education-real life situation
3. Higher Secondary Education-conservation
4. Higher Education-sustainable development

2. **Natural Resources** is the second disciplinary nature of environmental study.

The Environment and Resources are of most concern to the present society and also for future generations. The present society has to educate the future generations in the light of resources of the country which are to be utilised adequately.

The Natural Resources in the form of matter and energy are of vital significance for the successful survival of all types of life on the planet earth in general and for human being in particular. In fact, all aspects of human society(social,cultural,political and economic) depend on the resources.The very fundamental principle of resources is that resources even renewable are finite.For example,air and water are in abundance and are very much renewable but when they are polluted and degraded,they become unusable and non-renewable because it becomes very difficult to restore the original quality of air and water if they have been degraded beyond a certain limit.

The resources are fundamental base for the economic growth and development of human society but their withdrawal from the nature,mode of their uses by human being and their disposal have enormous adverse effect on the environment.Mishandling of resources and negligence of the upkeep of resouces also affect the environment adversely.

3. **Geographical nature** is the third disciplinary nature of environmental study.

The field and study of geography is Earth and Man.Earth includes physical,biological,social,cultural and economic components of the environment with the reference to man.Thus, geography is the only discipline that can pursue the study of environment and organisms.This is to say geography studies the biosphere(the interface of air,land,water)in totality of all components of biosphere

-abiotic and biotic,their characterstics and interrelationships.Being an integrating science geography synthesis all the elements and components of planet earth into one body and links the social sciences with natural sciences.

Geographers are the only scientists who can recognise and identify the environmental regions,locate them in space and present them on maps.The geographers recognise that the quality of life layer varies from place to place in terms of richness or poverty of life forms capable of being supported.Thus,geography discipline directly and widely concern with environment.

4. **Social Study** is the fourth disciplinary nature of environmental science.

The society is now awakened towards environment problems and the public concern about the quality of environment has reached.The enviornmental improvement measures being costly and long term investment of time,money and resources may ecllipse the public interest in the implementation of Environmental improvement of programmes.Environment changes through time and the whole sequence of changes is completed in five stages termed as "Issue Attention Cycle".

1) First stage : It is characterised by no public attention towards environmental problems except a few experts and interest groups who are seized with such problems.
2) Second stage : This is the stage of discovery and euporic enthusiasm--when the issue of environmental problems catches the public attention and the public are so alarmed and become so enthusiastic that they readily respond to solve the problems without having any care for the cost of investment.
3) Third stage : It is marked by realisation of the cost of significant progress That is to say that the technological development may not always be the best solution to the environmental problems.
4) Fourth stage : This is characterised by public interest towards environmental improvement programmes due to realisation ofv higher cost of solution environmental

problems and difficulties in implementing the environmental improvement programmes.

5) Fifth stage : It is marked by public interest in the issues of environmental concerns but the public interest occassionally surges.Thus the public interests in environmental problems occur in spasmodic manner and its directly concerns with environment.

Components of Environment

Environment is a broad concept to surround the whole range of diverse surroundings in which we perceive ,experience and react to events and changes.There are basic components of Environment which as follows-

Earth

Biosphere or Ecological framework always remains in a dynamic equillibrium. Air,water, soil,animals , plants and bacteria are all linked to each other or to their surroundings by some means or the other. This system is called Environment. Environment includes all those conditions which affect our life. Plants ,animals ,water,air and land are important parts of our environment because living being cannot survive without them.If environmental pollution keeps on increasing with the present rate, the very existence of the life on earth is endangered.

Land

Land is one of the most important components of our environment as the soil is vital for life. An inch of soil takes anything from 500 to 1,000 years to build. The soil erosion is a natural process, but in its undisturbed condition the existing plant cover of grasses, shrubs and trees not only check soil but also replenish soil.

As human activity disturbs this balance of soil and vegetation, soil erosion is accelerated, Pollution starts its existence suffers from one type of degradation or the other and comes under the definition of waste land, which means the productivity from this area is much less than what it should be.

Another about 27 millions hectares are degraded by floods, salinity and alkalinity. It is estimated that our national soil loss is about 6,000 million tons annually, and with this soil we loss about 6 million tons of nutrients.

The application of inorganic fertilizers to our crop-lands and using highly toxic insecticides and pesticides to control crop diseases, has the harmful effects of indiscriminate use of the toxic chemicals in agriculture over a long period is contaminating food and feed. It is absolutely essentially to be cautious with the modern technology in agriculture if we want to sustain to our crop production at a reasonable high level to feed the rising population.

Another problem is incorrect land use and this is mainly responsible for land degradation. Good agriculture land and forest are being destroyed for faulty irrigation practices or are being used for brick manufacture or for urban development scheme.

Water

Water is vital for all forms of life on this earth the main source of water which is roughly of the order of 400 million hectare metre per year. Nearly 50 percent of it percolates into the soil. About 30 percent of it runs as surface run off and another 20 per cent evaporates into the atmosphere.

Pure and clean water is necessary for healthy environment. We cannot afford to waste our valuable water and resources through faulty land and water management practices.

Main source of water which is roughly of the order of 400 million hectare meter per year. Nearly 50 per cent of it percolates into the soil. About 30 per cent of it runs as surface run off and another 20 per cent evaporates into the atmosphere. The responsibility for maintaining civil amenities like water supply, drainage, slum clearance, open spaces and streets in the cities rests with our municipal bodies.

Pollution of water may be by different sources. Soil erosion in the catchment of the rivers, streams and ponds leads to excessive sediments load thus polluting the water system. Streams and other water systems are also polluted by municipal waster and industrial effluent, particularly.

Similarly industries have taken up anticipation measures only haltingly causing severe pollution to our water bodies.

Forest

Forests influences local and regional climates besides their productive uses *viz.* supply of energy in the form of wood and timber for constructon purposes. The make the climate milder, help to ensure a continuous flow of clean water, reduce soil erosion, and reduces chances of floods and droughts.

The satellite imagery has indicated that the total extent of forest cover in 1986 was only 46 million hectare which is hardly about 14 per cent of the total land area as against the recorded area of about 23 per cent. We are losing about 1.5 million hectares of forest cover every year.

Air

Air is life itself but pollutants are being added into the atmosphere. Industries like Oil Refineries, Fertilisers, Paper and Cement etc. produce hydrogen sulphate, oxides of sulphur, organic vapour and dust. Steel plants emit metal fumes, fluorides, oxides of sulphur and particulates. All these are health hazards. The Central Board for Pollution Control and scientist form our state in collaboration with State Board of Pollution conducted a study and have found that the consumption of motor spirit containing **Lead** has increased by almost twice the anticipated rate of surrounding air.

It is necessary the every development programme/project should have in built provision for environmental protection (Pollution control) and should take into consideration the consequences of the project/programme on the Environment (Pollution load).

Environmental Education based on scientific studies and social system approach will generate awareness in the common masses as well as to goverment for better developmental planning.

We protect and conserve our natural resources of land, water, vegetation, and air, it will be impossible to meet the rising demands of the present day population without fore-closing the achievements.

Mining

Extraction of minerals for industrial development and better living standard is inevitable. Mining, however, has a serious disturbing effect on the ecology and enviroment leading to pollution. Generation of various pollutants, directly and indirectly are in built features of mining activity. Mining ruins the land, water, forests and air.

Air polluion in mining areas causes repiratory diseases and eye ailments. A fter the stady the indicated in that chronic exposure to sulphur dioxide and nitrous oxide has an effect on mortality rate of cancer. The developed countries like U.S.A., U.K. etc. there should be low requiring that reclamation of the mine area has, to be undertaken by the mining company and the land must be stored in original condition after mining

Classification of Environment

The Environment refers to all non-project related issues and, therefore, includes both natural, social and health aspects.Factors of the natural and social environment which have been considered are:

Atmospheric environment, specifically air quality, energy, light and noise.

Physico-chemical environment: climate, geology, geomorphology and topography, soils, sediments, hydrology and hydrodynamics of Bonny River, water chemistry,hydrogeology and surface water quality

Biological environment, considering the flora and fauna in four land and water systems pertinent to the natural environment of Bonny Island: the estuary and coastal system, mangrove forest, freshwater swamp forest and dry-land rainforest and associated farmland.

Social, cultural and economic environment, considering both: socio-economic aspects - including demography, economic activity, infrastructure, exploitation of the natural environment and socio-cultural aspects - including ethnology, religion, education, social structure, social concerns, history and nature conservation.

Health Environment, considering the general health status of the population in the area, existing workforce, their families and influx workforce/migrant people directly or indirectly associated/ attracted by the project; health care statistics - vital statistics and disease patterns, and health infrastructure - perceptions about health and health needs.

Atmosphere-btructure and its Function

Among the nine planets of the solar system,earth is the only one planet where life exists.Plants and animals exists on earth because it has all the favourable conditions required for survival of life.The part of the earth where plants and animals exists is called Biosphere.Biosphere is further divided into three parts: Lithosphere,Hydrosphere and Atmosphere. Lithosphere is that part of the earth rocks ,soil,plants and animals are found. The part of the earth where water is present is called Hydrosphere.All the animals and plants need water to live. The blanket of air which surrounds the earth up to a height of about 100 kms is called Atmosphere.

Plants and animals exists on earth because it has all the favourable conditions required for survival of life.

Even on earth life is not found every where because survival requires a particular type of environment. The blanket of air which surrounds the earth up to a height of about 100 km is called atmosphere. Atmpsphere is a balanced mixture of several gases like nitrogen, oxygen and carbon dioxide. The major atmosphere extends upto 320 km above lithosphere and hydrosphere. About 10 km below and 10 km above the surface of the earth is biosphere. In reality 90 per cent living beings live in the area between 1 km below and 1 km above the earth's surface. Biosphere or ecological framework always remains in a dynamic equilibrium. Air, water, soil, animals, plants and bacteria are all linked to each other or to their surroundings by some means or the other. This system is called *Environment.* Environment includes all those conditions which affect our life. Plants, animals, living being cannot survive without them.

The atmosphere forms an insulating blanket around the earth. Without it the temperature at the equator would rise to 180 during the day and drop as low as—220°F at night. It burns up meteors that would bombard the surface of the earth from space. Without the atmosphere there would be no sound and no flight. There would be no conventional longdistance radio communication, for this is dependent on the electrons in the upper atmosphere. Without air there would be no lightning, no clouds, no wind, no rain, no snow, and no fire.

The atmosphere shields of the earth from lethal concentrations of ultraviolet radiation. It selectively filters the sun's rays so that only two small segments of the electromagnetic spectrum penetrate to the earth's surface in appreciable amounts.

One segment is the *optical window,* consisting essentially of the visible spectrum of light, from near-near ultraviolet to near-infrared. The other is the radio window, consisting of radio waves from about 1 centimeter to 40 meters in length. It is impossible to define the limits of the atmosphere because the atmosphere becomes progressively tendious with increasing distance from the earth. There is no boundary between the atmosphere and the void of outer space.

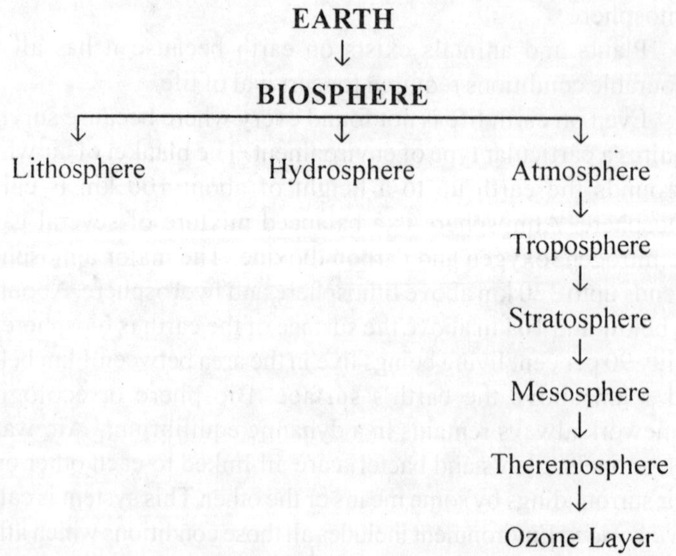

The Zones of the Atmosphere

The temperature of the atmosphere changes with increasing altitude, scientists have defined zones of the atmosphere. Each zone is more or less a spherical shell of the atmosphere.

Troposphere

The zone nearest the earth, the troposphere, is named from the *Greek tropein* (to turn, to rotate, to change). One property that changes in the troposphere is the temperature. It drops to—55°C (—65°F) by the time an altitude of about 16-18 km (10-11 miles) is reached. The rate of temperature drop is about 6.4°C per 1000 meters (3.5°F per 1000 feet), a figure called the normal environmental lapse ratc. The region through which this lapse rate occurs defines the troposphere.

The jet stream occurs at an altitude of 10.55-15 kilometers (35,000-40,000 feet; 6.5-7.5 miles).

Stratosphere

This zone starts at the top of the troposphere, at a thin atmospheric "shell" called the tropopause, where the temperature begins to stay fairly constant, increasing slightly, in going to an altitude of 50 kilometers (31 miles). The stratosphere includes much of the *ozone layer.* The warming trend in the stratosphere is caused by a cycle of chemical changes in the ozone layer which converts all the incoming high energy ultraviolet rays to heat. If these rays reached the earth's surface life as we know it would be impossible. The *stratopause* marks the narrow zone at the top of the stratosphere where the temperature begins to fall again with increasing altitude as the mesosphere is entered.

Mesosphere

The mesosphere, on top of the stratosphere, extends roughly to 80 kilometeres (50 miles), and its temperature drops to between—80 and–90°C at the mesopause, a thin zone where the temperature stalizes and soon starts to rise again as the thermosphere is ascended.

Theremosphere

This zone is above 80 kilometers (50 mites) altitude. The temperature becomes very high but the air has such a low density that it can hold very little heat. Any, denser object is the thermosphere will be extremely hot in sunlight bt very cold at night.

Changes in chemical composition or activity, also occur with altitude.

Above the homosphere lies the *heterosphere.* For the first 40 kilometers (25 miles), the ionosphere, the very thin concentration of atoms and molecules exists largely as electrically charged particles called ions. This globe enveloping band of ions reflects outgoing radio waves back to earth, making radio transmission possible well beyond the horizon. Television waves are not reflected, and a TV transmitter's range does not extent beyond the horizone Meteorites entering the earth's atmosphere generally out in the ionsphere.

Earth's Atmospheres

The Earth's atmosphere (or air) is a layer of gases surrounding the planet Earth that is retained by the Earth's gravity. Dry air contains roughly (by volume) 78.08% nitrogen, 20.95% oxygen, 0.93% argon, 0.038% carbon dioxide, and trace amounts of other gases. Air also contains a variable amount of water vapor, on average around 1%. The atmosphere protects life on Earth by absorbing ultraviolet solar radiation, warming the surface through heat retention (greenhouse effect), and reducing temperature extremes between day and night.

There is no definite boundary between the atmosphere and outer space. It slowly becomes thinner and fades into space. An altitude of 120 km (75 mi) marks the boundary where atmospheric effects become noticeable.

The temperature of the Earth's atmosphere varies among five different atmospheric layers (ordered highest to lowest, the ionosphere is part of the thermosphere):

Exosphere

From 500-1,000 km (310-620 mi) up to 10,000 km (6,200 mi),

contain free-moving particles that may migrate into and out of the magnetosphere or the solar wind.

Exobase

Also known as the critical level, it is the lower boundary of the exosphere.

Ionosphere

The part of the atmosphere that is ionized by solar radiation stretches from 50 to 1,000 km (31 to 620 mi) and typically overlaps both the exosphere and the thermosphere. It plays an important part in atmospheric electricity and forms the inner edge of the magnetosphere. Because of it's charged particles, it has practical importance because it influences, for example, radio propagation on the Earth.

Thermopause

The boundary above the thermosphere, it varies in height from 500-1,000 km (310-620 mi).

Thermosphere

From 80-85 km (50-53 mi; 260,000-280,000 ft) to over 640 km (400 mi; 2,100,000 ft), temperature increasing with height. Although the temperature can rise to 1,500 °C (2,730 °F), a person would not feel warm because of the extreme low pressure. The International Space Station orbits in this layer, between 320 and 380 km (200 and 240 mi).

Mesopause

The temperature minimum at the boundary between the thermosphere and the mesosphere. It is the coldest place on Earth, with a temperature of 100 °C (148.0 °F; 173.1 K).

Mesosphere "Meso"meaning middle. The mesosphere extends from about 50 km (31 mi; 160,000 ft) to the range of 80-85 km (50-53 mi; 260,000-280,000 ft). Temperature decreases with height, reaching 100 °C (148.0 °F; 173.1 K) in the upper mesosphere. This is also where most meteors burn up when entering the atmosphere.

Stratopause

The boundary between the mesosphere and the stratosphere, typically 50 to 55 km (31 to 34 mi; 160,000 to 180,000 ft). The pressure here is 1/1000th sea level.

Stratosphere

From the Latin word "stratus" meaning spreading out. The stratosphere extends from the troposphere's 7-17 km (4.3-11 mi; 23,000-56,000 ft) range to about 51 km (32 mi; 170,000 ft). Temperature increases with height. The stratosphere contains the ozone layer, the part of the Earth's atmosphere which contains relatively high concentrations of ozone. "Relatively high" means a few parts per million-much higher than the concentrations in the lower atmosphere but still small compared to the main components of the atmosphere. It is mainly located in the lower portion of the stratosphere from approximately 15-35 km (9.3-22 mi; 49,000-110,000 ft) above Earth's surface, though the thickness varies seasonally and geographically.

Ozone Layer

Though part of the Stratosphere, the ozone layer is considered as a layer of the Earth in itself due to the fact that its physical and chemical composition is far different to the Stratosphere. Ozone in the earth's stratosphere is created by ultraviolet light striking oxygen molecules containing two oxygen atoms (O^2), splitting them into individual oxygen atoms (atomic oxygen); the atomic oxygen then combines with unbroken O^2 to create ozone, O^3. The ozone molecule is also unstable (although, in the stratosphere, long-lived) and when ultraviolet light hits ozone it splits into a molecule of O^2 and an atom of atomic oxygen, a continuing process called the ozone-oxygen cycle, thus creating an ozone layer in the stratosphere, the region from about 10 to 50 km (32,000 to 164,000 feet) above Earth's surface. About 90% of the ozone in our atmosphere is contained in the stratosphere. Ozone concentrations are greatest between about 20 and 40 km, where they range from about 2 to 8 parts per million.

Tropopause

The boundary between the stratosphere and troposphere.

Troposphere—The troposphere is the lowest layer of the atmosphere,it is the zone which is nearest the earth; it is named from the Greek tropein (to turn,to rotate,to change).One property that changes in the troposphere is the temperature.It begins at the surface and extends to between 7 km (23,000 ft) at the poles and 17 km (56,000 ft) at the equator, with some variation due to weather factors. The troposphere has a great deal of vertical mixing because of solar heating at the area. This heating makes air masses less dense so they rise. When an air mass rises, the pressure upon it decreases so it expands, doing work against the opposing pressure of the surrounding air. To do work is to expend energy, so the temperature of the air mass decreases. As the temperature decreases, water vapor in the air mass may condense or solidify, releasing latent heat that further uplifts the air mass. This process determines the maximum rate of decline of temperature with height, called the adiabatic lapse rate or normal environmental lapse rate. The troposphere contains roughly 80% of the total mass of the atmosphere. Fifty percent of the total mass of the atmosphere is located in the lower 5.6 km (18,000 ft) of the troposphere.

The average temperature of the atmosphere at the surface of Earth is 20 °C (68 °F; 293 K).

Compositions

Composition of Earth's atmosphere. The lower pie represents the least common gases that compose 0.038% of the atmosphere. Values normalized for illustration.

Mean atmospheric water vapor	*Composition of dry atmosphere, by volume ppmv: parts per million by volume*	*Gas Volume*
Nitrogen (N_2)	780,840 ppmv	(78.084%)
Oxygen (O_2)	209,460 ppmv	(20.946%)
Argon (Ar)	9,340 ppmv	(0.9340%)
Carbon dioxide (CO_2)	383 ppmv	(0.0383%)
Neon (Ne)	18.18 ppmv	(0.001818%)
Helium (He)	5.24 ppmv	(0.000524%)
Methane (CH_4)	1.745 ppmv	(0.0001745%)
Krypton (Kr)	1.14 ppmv	(0.000114%)
Hydrogen (H_2)	0.55 ppmv	(0.000055%)
Nitrous oxide (N_2O)	0.3 ppmv	(0.00003%)
Xenon (Xe)	0.09 ppmv	(9x10-6%)
Ozone (O_3)	0.0 to 0.07 ppmv	(0%-7x10-6%)
Nitrogen dioxide (NO_2)	0.02 ppmv	(2x10-6%)
Iodine (I)	0.01 ppmv	(1x10-6%)
Carbon monoxide (CO)	trace	
Ammonia (NH_3)	trace	
Not included in above dry atmosphere:		
Water vapor (H_2O)		
0.40% over full atmosphere, typically 1%-4% at surface		

Biosphere

The biosphere is the biological component of earth systems, which also include the lithosphere, hydrosphere, atmosphere .The biosphere includes all living organisms on earth, together with the dead organic matter produced by them.

The biosphere is simply "life on Earth"-the sum total, that is, of all living things on Earth. Yet the whole is more than the sum of the parts: not only is the biosphere an integrated system whose many components fit together in complex ways, but it also works, in turn, in concert with the other major earth systems. The latter include the geosphere, hydrosphere, and atmosphere, through which circulate the chemical elements and compounds essential to life. Among these elements is carbon, a part of all living things, which also cycles through the nonliving realms of soil, water, and air-just one of many vital biogeo-chemical cycles. As for the compounds on which life depends, none is more important than water, which, though it is the focal point of the hydrosphere, passes through the various earth systems as well.

Organisms participate in the hydrologic cycle by providing moisture to the air through the process of transpiration, and they likewise benefit from the downward movement of moisture in the form of precipitation. These and many other interactions make it easy to see why scientists speak of Earth as a system-and why some go even further and call it a living thing.

Hydrosphere

Approximately 74% of the Earth's surface is covered by water, in either the liquid or solid state. These waters, combined with minor contributions from ground waters, constitute the hydrosphere.

The oceans account for about 97% of the weight of the hydrosphere, while the amount of ice reflects the Earth's climate, being higher during periods of glaciation. There is a considerable amount of water vapor in the atmosphere. The circulation of the waters of the hydrosphere results in the weathering of the landmasses. The annual evaporation from the world oceans and from land areas results in an annual precipitation of 320,000km3 (76,000 mi3) on the world oceans and 100,000 km3 (24,000 mi3) on land areas. The rainwater falling on the continents, partly taken up by the ground and partly by the streams, acts as an erosive agent before returning to the seas.

The unique chemical properties of water make it an effective solvent for many gases, salts, and organic compounds. Circulation of water and the dissolved material it contains is a highly dynamic process driven by energy from the Sun and the interior of the Earth. Each component has its own geochemical cycle or pathway through the hydrosphere, reflecting the component's relative abundance, chemical properties, and utilization by organisms. The introduction of materials by humans has significantly altered the composition and environmental properties of many natural waters.

Lithosphere

The rigid or mechanically strong outer layer of the Earth that can support stress. The lithosphere is divided into 12 major plates, the boundaries of which are zones of intense activity that produce many of the large-scale geological features that characterize the Earth. These plates move as coherent units with velocities of up to several centimeters per year, and their relative movement and interaction form the foundation for the theory of plate tectonics.

The lithosphere comprises the crust (either continental or oceanic) and a portion of the upper mantle that together overlie a zone of relative weakness termed the asthenosphere. The boundary between the crust and the mantle is known as the Mohorovii discontinuity (Moho), and is compositional in origin-that is, the crust and mantle are distinguished by fundamental differences in rock chemistry.

In contrast, the boundary between the lithosphere and the asthenosphere represents an isotherm that separates a conductively cooling lithosphere from a quasi-isothermal convecting asthenosphere. The asthenosphere differs from the overlying lithosphere principally in its ability to flow on geological time scales. These differences arise from the fact that temperature (and thus the fluid or flow properties of rocks) increases as a function of depth in the Earth. Whereas the lithosphere tends to be resistant to deformation, the asthenosphere deforms by flowing. The lithosphere is either oceanic or continental, each type being fundamentally different in terms of the formation and composition of the rocks that constitute the crust and upper mantle. The chemical lithosphere is defined as a chemical boundary layer between the surface of the Earth and the asthenosphere that cools by conduction and contains both the material differentiated or extracted from the mantle.

2

THE ECOSYSTEM

An **Ecosystem** consists of the biological community that occurs in some locale, and the physical and chemical factors that make up its non-living or abiotic environment. There are many examples of ecosystems — a pond, a forest, an estuary, a grassland. The boundaries are not fixed in any objective way, although sometimes they seem obvious, as with the shoreline of a small pond. Usually the boundaries are chosen for practical reasons having to do with the goals of the particular study.

The study of ecosystems mainly consists of the study of certain processes that link the living, or biotic, components to the non-living, or abiotic, components. Energy transformations and biogeochemical cycling are the main processes that comprise the field of ecosystem ecology.

Ecology generally is defined as the interactions of organisms with one another and with the environment in which they occur. We can study ecology at the level of the individual, the population, the community, and the ecosystem. Studies of populations usually focus on the habitat and resource needs of individual species, their behaviours, population growth, and what limits abundance. Studies of communities examine how populations of many species interact with one another, such as predators and their prey, or competitors that share common needs.

An **Ecosystem** was defined as a dynamic entity composed of a biological community and its associated abiotic environment. Often the dynamic interactions that occur within an ecosystem are numerous and complex. Ecosystems are also always undergoing alterations to their biotic and abiotic components. Some of these alterations begin first with a change in the state of one component of the ecosystem which then cascades and sometimes amplifies into other components because of relationships.

In recent years, the impact of humans has caused a number of dramatic changes to a variety of ecosystems found on the Earth. Humans use and modify natural ecosystems through agriculture, forestry, recreation, urbanization, and industry. The most obvious impact of humans on ecosystems is the loss of biodiversity. The number of extinctions caused by human domination of ecosystems has been steadily increasing since the start of the Industrial Revolution. The frequency of species extinctions is correlated to the size of human population on the Earth which is directly related to resource consumption, land-use change, and environmental degradation. Other human impacts to ecosystems include species invasions to new habitats, changes to the abundance and dominance of species in communities, modification of biogeochemical cycles, modification of hydrologic cycling, pollution, and climatic change.

Major Components of Ecosystems

Ecosystems are composed of a variety of abiotic and biotic components that function in an interrelated fashion. Some of the more important components are: ***soil***, ***atmosphere***, ***radiation from the Sun***, ***water***, and ***living organisms***.

Soils are much more complex than simple sediments. They contain a mixture of weathered rock fragments, highly altered soil mineral particles, **organic matter**, and living organisms. Soils provide **nutrients**, water, a home, and a structural growing medium for organisms. The vegetation found growing on top of a soil is closely linked to this component of an ecosystem through nutrient cycling.

The *atmosphere* provides organisms found within ecosystems with carbon dioxide for **photosynthesis** and oxygen for **respiration**. The processes of **evaporation**, **transpiration**, and **precipitation** cycle water between the atmosphere and the Earth's surface.

Solar radiation is used in ecosystems to heat the atmosphere and to **evaporate** and **transpire** water into the atmosphere. Sunlight is also necessary for **photosynthesis**. Photosynthesis provides the energy for plant growth and metabolism, and the organic food for other forms of life.

Most living tissue is composed of a very high percentage of *water*, up to and even exceeding 90%. The **protoplasm** of a very few cells can survive if their water content drops below 10%, and most are killed if it is less than 30-50%. Water is the medium by which mineral nutrients enter and are translocated in plants. It is also necessary for the maintenance of leaf turgidity and is required for photosynthetic chemical reactions. Plants and animals receive their water from the Earth's surface and soil. The original source of this water is precipitation from the atmosphere.

Ecosystems are composed of a variety of ***living organisms*** that can be classified as **producers**, **consumers**, or **decomposers**. **Producers** or **autotrophs**, are organisms that can manufacture the organic compounds they use as sources of energy and **nutrients**. Most producers are green plants that can manufacture their food through the process of **photosynthesis**. **Consumers** or **heterotrophs** get their energy and nutrients by feeding directly or indirectly on producers. We can distinguish two main types of consumers. **Herbivores** are consumers that eat plants for their energy and nutrients. Organisms that feed on herbivores are called **carnivores**. Carnivores can also consume other carnivores. Plants and animals supply organic matter to the soil system through shed tissues and death. Consumer organisms that feed on this organic matter, or **detritus**, are known as **detritivores** or **decomposers**. The organic matter that is consumed by the detritivores is eventually converted back into **inorganic** nutrients in the soil. These nutrients can then be used by plants for the production of organic compounds.

The following **graphical** model describes the major ecosystem components and their interrelationships:—

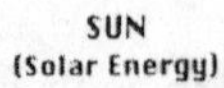

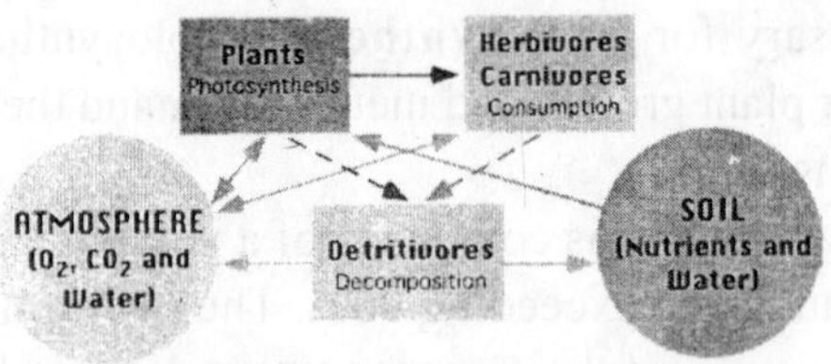

Relationship Within Ecosystem

Structure of Ecosystem

This figure with the plants, zebra, lion, and so forth illustrates the two main ideas about how ecosystems function: ecosystems have energy flows and ecosystems cycle materials. These two processes are linked, but they are not quite the same (see Figure 1).

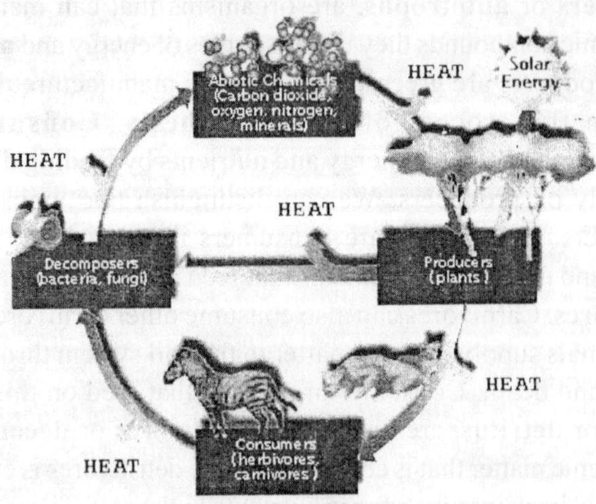

Figure 1. Energy flows and material cycles.

Energy enters the biological system as light energy, or photons, is transformed into chemical energy in organic molecules by cellular processes including photosynthesis and respiration, and ultimately is converted to heat energy. This energy is dissipated, meaning it is lost to the system as heat; once it is lost it cannot be recycled. Without the continued input of solar energy, biological systems would quickly shut down. Thus the earth is an open system with respect to energy.

Elements such as carbon, nitrogen, or phosphorus enter living organisms in a variety of ways. Plants obtain elements from the surrounding atmosphere, water, or soils. Animals may also obtain elements directly from the physical environment, but usually they obtain these mainly as a consequence of consuming other organisms. These materials are transformed biochemically within the bodies of organisms, but sooner or later, due to excretion or decomposition, they are returned to an inorganic state. Often bacteria complete this process, through the process called decomposition or mineralization.

During decomposition these materials are not destroyed or lost, so the earth is a closed system with respect to elements (with the exception of a meteorite entering the system now and then). The elements are cycled endlessly between their biotic and abiotic states within ecosystems. Those elements whose supply tends to limit biological activity are called nutrients.

Food Chains

Food chains, also called, **food networks** and/or **trophic social networks**, describe the eating relationships between species within an ecosystem. Organisms are connected to the organisms they consume by lines representing the direction of organism or energy transfer. It also shows how the energy from the producer is given to the consumer. Typically a food chain or food web refers to a graph where only connections are recorded, and a food network or ecosystem network refers to a network where the connections are given weights representing the quantity of nutrients or energy being transferred.

Organisms represented in food chains

Primary producer, commonly forming autotrophs, produce simple organic substances (essentially "food") from an energy source and inorganic materials. These organisms are typically photosynthetic. Organisms that get their energy by organic substances are called heterotrophs. Heterotrophs include herbivores, which obtain their energy by consuming live plants; carnivores, which obtain energy from eating live animals; as well as detritivores, scavengers and decomposers, which all consume dead biomass. Energy enters the food chain from the sun. Some energy and/or biomass is lost at each stage of the food chain as; feces (solid waste), movement energy and heat energy (especially by warm-blooded creatures). Therefore, only a small amount of energy and biomass is incorporated into the consumer's body and transferred to the next feeding level, thus showing a Pyramid of Biom

Flow of food chains

A **food chain** is the flow of energy from one organism to the next and to the next and so on. Organisms in a food chain are grouped into trophic levels, based on how many links they are removed from the primary producers. Trophic levels may contain either a single species or a group of species that are presumed to share both predators and prey. They usually start with a plant and end with a carnivore. The diagram below is a *food chain* from a Swedish lake. Osprey feed on northern pike that feed on perch that eat bleak that feed on freshwater shrimp. Though unshown, the primary producers of this food chain are probably autotrophic phytoplankton. Phytoplankton and algae form the base of most freshwater food chains. It is often the case that biomass of each trophic level decreases from the base of the chain to the top. This is because energy is lost to the environment with each transfer. On average, only 10% of the organism's energy is passed on to its predator. The other 90% is used for the organism's life processes or it is lost as heat to the environment. Graphic representations of the biomass or productivity at each tropic level are called trophic pyramids. In this food chain for example, the biomass of osprey is smaller than the biomass of pike, which is smaller than the biomass of perch. Some producers, especially phytoplankton, are so productive and have such a high turnover rate that they can actually support a larger biomass of grazers. This is called an *inverted pyramid*, and can occur when consumers live longer and grow more slowly than the organisms they consume. In this food chain, the productivity of phytoplankton is much greater than that of the zooplankton consuming them. The biomass of the phytoplankton, however, may actually be less than that of the copepods. Directly linked to this are pyramids of numbers, which show that as the chain is traveled along, the number of consumers at each level drops very significantly, so that a single top consumer (e.g. a Polar Bear) will be supported by literally millions of separate producers (e.g. Phytoplankton).

Example of a *food chain* in a Swedish lake

The source of all food is the activity of autotrophs, mainly photosynthesis by plants.

- They are called **producers** because only they can manufacture food from inorganic raw materials.
- This food feeds **herbivores**, called **primary consumers**.
- **Carnivores** that feed on herbivores are called **secondary consumers**.
- Carnivores that feed on other carnivores are **tertiary** (or higher) consumers.

Such a path of food consumption is called a **food chain**.

Each level of consumption in a food chain is called a **trophic level**.

The table gives one example of a food chain and the trophic levels represented in it.

Grass →	Grasshopper →	Toad →	Snake →	Hawk →	Bacteria of decay
In general,					
Autotrophs (Producers) →	Herbivores (Primary Consumers) →	Carnivores (Secondary, tertiary, etc. consumers) →			Decomposers

Food Webs

Most food chains are interconnected. Animals typically consume a varied diet and, in turn, serve as food for a variety of other creatures that prey on them. These interconnections create **food webs**.

The real world, of course, is more complicated than a simple food chain. While many organisms do specialize in their diets (anteaters come to mind as a specialist), other organisms do not. Hawks don't limit their diets to snakes, snakes eat things other than mice, mice eat grass as well as grasshoppers, and so on. A more realistic depiction of who eats whom is called a food web; an example is shown below:

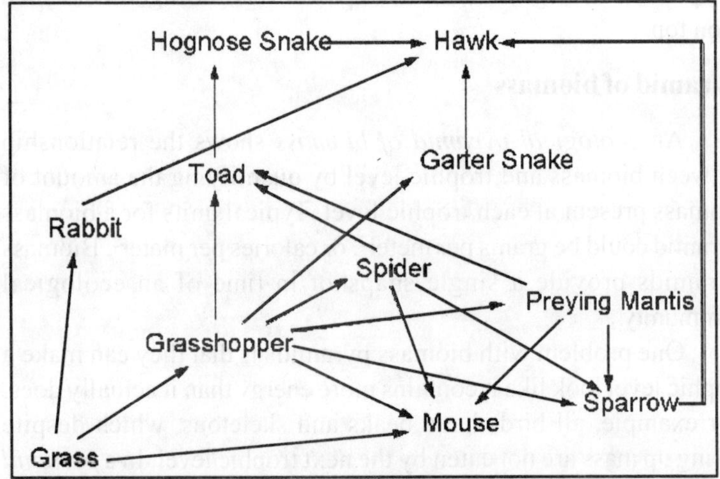

It is when we have a picture of a food web in front of us that the definition of food chain makes more sense. We can now see that a food web consists of interlocking food chains, and that the only way to untangle the chains is to trace *back* along a given food chain to its source.

The food webs you see here are *grazing food chains* since at their base are producers which the herbivores then graze on. While grazing food chains are important, in nature they are outnumbered by *detritus-based food chains*. In detritus-based food chains, decomposers are at the base of the food chain, and sustain the carnivores which feed on them. In terms of the weight (or biomass) of animals in many ecosystems, more of their body mass can be traced back to detritus than to living producers.

Ecological Pyramids

An **ecological pyramid** (or **trophic pyramid**) is a graphical representation designed to show the biomass or productivity at each trophic level in a given ecosystem. ***Biomass pyramids*** show the abundance or biomass of organisms at each trophic level, while **Productivity** ***pyramids*** show the production or turn-over in biomass. *Ecological pyramids* begin with producers on the bottom and proceed through the various trophic levels, the highest of which is on top.

Pyramid of biomass

An *ecological pyramid of biomass* shows the relationship between biomass and trophic level by quantifying the amount of biomass present at each trophic level. Typical units for a biomass pyramid could be grams per meter^2, or calories per meter^2. Biomass pyramids provide a single snapshot in time of an ecological community.

One problem with biomass pyramids is that they can make a trophic level look like it contains more energy than it actually does. For example, all birds have beaks and skeletons, which despite taking up mass are not eaten by the next trophic level. In a *pyramid of biomass* the skeletons and beaks would still be quantified even though they do not contribute to the overall flow of energy into the next trophic level.

Also, the pyramid of biomass may be 'inverted'. For example, in a pond ecosystem, the standing crop of phytoplankton, the major producers, at any given point will be lower than the mass of the heterotrophs, such as fish and insects. This is explained as the phytoplankton reproduce very quickly.

Pyramid of productivity

An *ecological pyramid of productivity* is often more useful, showing the production or turnover of biomass at each trophic level. Instead of showing a single snapshot in time, productivity pyramids show the flow of energy through the food chain. Typical units would be grams per meter^2 per year or calories per meter^2 per year. As with the others, this graph begins with producers at the bottom and places higher trophic levels on top.

When an ecosystem is healthy, this graph generally looks like the standard *ecological pyramid*. This is because in order for the ecosystem to sustain itself, there must be more energy at lower trophic levels than there is at higher trophic levels. This allows for organisms on the lower levels to maintain a stable population, but to also feed the organisms on higher trophic levels, thus transferring energy up the pyramid. The exception to this generalization is when portions of a food web are supported by inputs of resources from outside of the local community. In small, forested streams, for example, many consumers feed on dead leaves which fall into the stream. The productivity at the second trophic level is therefore greater than could be supported by the local primary production.

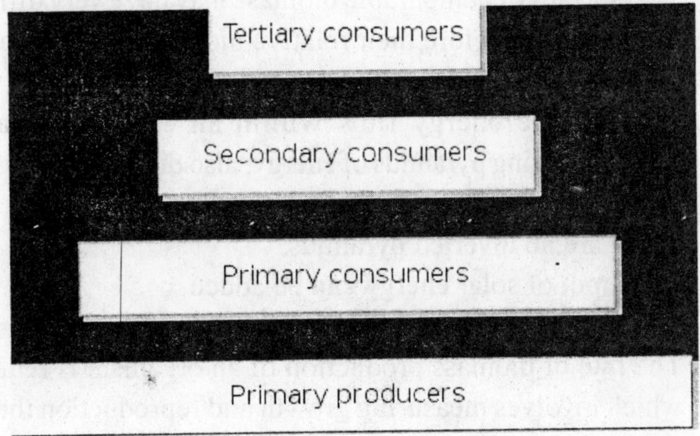

An ecological pyramid

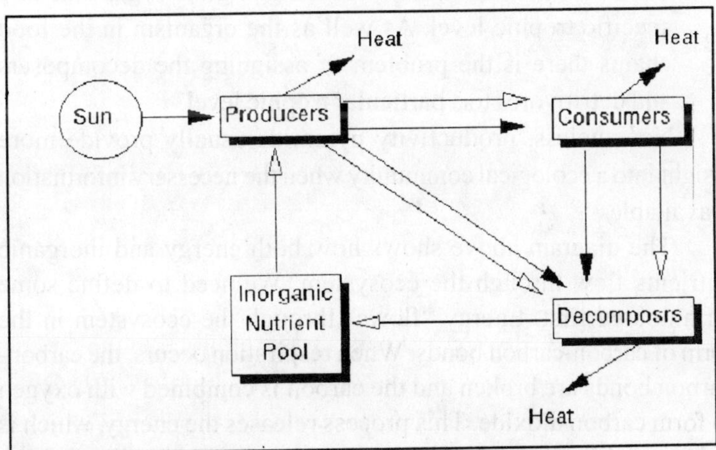

Energy Flow Through the Ecosystem

When energy is transferred to the next trophic level, typically only 10% of it is used to build new biomass, becoming stored energy (the rest going to metabolic processes). As such, in a *pyramid of productivity* each step will be 10% the size of the previous step (100, 10, 1, 0.1, 0.01, 0.001 etc.).

The advantages of the *pyramid of productivity*:

- It takes account of the rate of production over a period of time.
- Two species of comparable biomass may have very different life spans. Therefore their relative biomasses is misleading, but their productivity is directly comparable.
- The relative energy flow within an ecosystem can be compared using pyramids of energy; also different ecosystems can be compared.
- There are no inverted pyramids.
- The input of solar energy can be added.

The disadvantages of the *pyramid of productivity*:

- The rate of biomass production of an organism is required, which involves measuring growth and reproduction through time.
- There is still the difficulty of assigning the organisms to a specific trophic level. As well as the organism in the food chains there is the problem of assigning the decomposers and detritivores to a particular trophic level.

Nonetheless, productivity pyramids usually provide more insight into a ecological community when the necessary information is available.

The diagram above shows how both energy and inorganic nutrients flow through the ecosystem. We need to define some terminology first. Energy "flows" through the ecosystem in the form of carbon-carbon bonds. When respiration occurs, the carbon-carbon bonds are broken and the carbon is combined with oxygen to form carbon dioxide. This process releases the energy, which is either used by the organism (to move its muscles, digest food, excrete wastes, think, etc.) or the energy may be lost as heat. The dark arrows represent the movement of this energy. Note that all energy comes from the sun, and that the ultimate fate of all energy in ecosystems is to be lost as heat. Energy does not recycle!!

The other component shown in the diagram are the inorganic nutrients. They are inorganic because they do not contain carbon-carbon bonds. These inorganic nutrients include the phosphorous in your teeth, bones, and cellular membranes; the nitrogen in your amino acids (the building blocks of protein); and the iron in your blood (to name just a few of the inorganic nutrients). The movement of the inorganic nutrients is represented by the open arrows. Note that the autotrophs obtain these inorganic nutrients from the inorganic nutrient pool, which is usually the soil or water surrounding the plants or algac. These inorganic nutrients are passed from organism to organism as one organism is consumed by another. Ultimately, all organisms die and become detritus, food for the decomposers. At this stage, the last of the energy is extracted (and lost as heat) and the inorganic nutrients are returned to the soil or water to be taken up again. The inorganic nutrients are recycled, the energy is not.

Many of us, when we hear the word "nutrient" immediately think of calories and the carbon-carbon bonds that hold the caloric energy. When writing about energy flow and inorganic nutrient flow in an ecosystem, you must be clear as to what you are referring. Unmodified by "inorganic" or "organic", the word "nutrient" can leave your reader unsure of what you mean. This is one case in which the scientific meaning of a word is very dependent on its context. Another example would be the word "respiration", which to the layperson usually refers to "breathing", but which means "the extraction of energy from carbon-carbon bonds at the cellular level" to most scientists (except those scientists studying breathing, who use respiration in the lay sense).

To summarize: In the flow of energy and inorganic nutrients through the ecosystem, a few generalizations can be made:

1. The ultimate source of energy (for most ecosystems) is the sun
2. The ultimate fate of energy in ecosystems is for it to be lost as heat.
3. Energy and nutrients are passed from organism to organism through the food chain as one organism eats another.
4. Decomposers remove the last energy from the remains of organisms.
5. Inorganic nutrients are cycled, energy is not.

Biogeochemistry

The term ***Biogeochemistry*** is defined as the study of how living systems influence, and are controlled by, the geology and chemistry of the earth. Thus biogeochemistry encompasses many aspects of the abiotic and biotic world that we live in.

There are several main ***principles and tools*** that biogeochemists use to study earth systems. Most of the major environmental problems that we face in our world today can be analyzed using biogeochemical principles and tools. These problems include global warming, acid rain, environmental pollution, and increasing greenhouse gases. The principles and tools that we use can be broken down into 3 major components: ***element ratios, mass balance, and element cycling.***

1. Element ratios–In biological systems, we refer to important elements as ***"conservative"***. These elements are often nutrients. By "conservative" we mean that an organism can change only slightly the amount of these elements in their tissues if they are to remain in good health. It is easiest to think of these conservative elements in relation to other important elements in the organism. For example, in healthy algae the elements C, N, P, and Fe have the following ratio, called the ***Redfield ratio*** after the oceanographer who discovered it:

C : N : P : Fe = 106 : 16 : 1 : 0.01

Once we know these ratios, we can compare them to the ratios that we measure in a sample of algae to determine if the algae are lacking in one of these limiting nutrients.

2. Mass Balance–Another important tool that biogeochemists use is a simple mass balance equation to describe the state of a system. The system could be a snake, a tree, a lake, or the entire globe. Using a mass balance approach we can determine whether the system is changing and how fast it is changing. The equation is:

Net Change = Input + Output + Internal Change

In this equation the net change in the system from one time period to another is determined by what the inputs are, what the outputs are, and what the internal change in the system was. The example given in class is of the acidification of a lake, considering the inputs and outputs and internal change of acid in the lake.

3. Element Cycling–Element cycling describes where and how fast elements move in a system. There are two general classes of systems that we can analyze, as mentioned above: ***closed and open systems.***

*A **closed system** refers to a system where the inputs and outputs are negligible compared to the internal changes. Examples of such systems would include a bottle, or our entire globe. There are two ways we can describe the cycling of materials within this closed system, either by looking at the rate of movement or at the pathways of movement.*

*1. **Rate** = number of cycles / time * as rate increases, productivity increases*

*2. **Pathways**-important because of different reactions that may occur*

*In an **open system** there are inputs and outputs as well as the internal cycling. Thus we can describe the rates of movement and the pathways, just as we did for the closed system, but we can also define a new concept called the **residence time**. The residence time indicates how long on average an element remains within the system before leaving the system.*

1. Rate

2. Pathways

3. Residence time, Rt

***Rt** = total amount of matter / output rate of matter*

(Note that the "units" in this calculation must cancel properly)

BioGeoChemical Cycles

We have already seen that while energy does not cycle through an ecosystem, chemicals do. The inorganic nutrients cycle through more than the organisms, however, they also enter into the atmosphere, the oceans, and even rocks. Since these *chem*icals cycle through both the *bio*logical and the *geo*logical world, we call the overall cycles biogeochemical cycles. Each chemical has its own unique cycle, but all of the cycles do have some things in common. *Reservoirs* are those parts of the cycle where the chemical is held in large quantities for long periods of time. In *exchange pools*, on the other hand, the chemical is held for only a short time. The length of time a chemical is held in an exchange pool or a reservoir is termed its *residence* time. The oceans are a reservoir for water, while a cloud is an exchange pool. Water may reside in an ocean for thousands of years, but in a cloud for a few days at best. The biotic community includes all living organisms. This community may serve as an exchange pool (although for some chemicals like carbon, bound in a sequoia for a thousand years, it may seem more like a reservoir), and also serve to move chemicals from one stage of the cycle to another. For instance, the trees of the tropical rain forest bring water up from the forest floor to be evaporated into the atmosphere. Likewise, coral endosymbionts take carbon from the water and turn it into limestone rock. The energy for most of the transportation of chemicals from one place to another is provided either by the sun or by the heat released from the mantle and core of the Earth.

While all inorganic nutrients cycle, we will focus on only 4 of the most important cycles - water, carbon ,oxygen, nitrogen, and phosphorous.

The Water Cycle

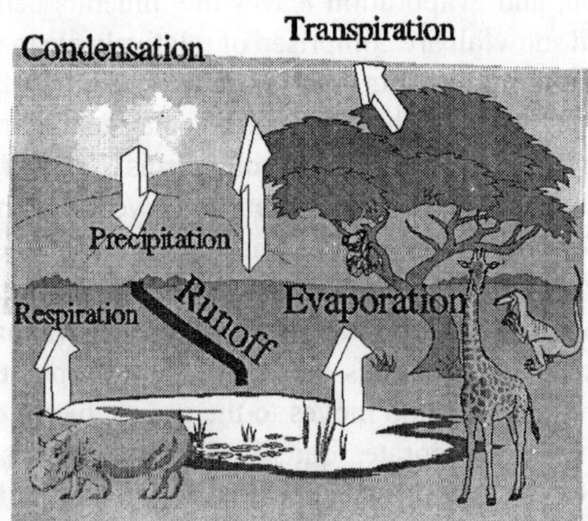

The Water Cycle

Key Features: In the water cycle, energy is supplied by the sun, which drives evaporation whether it be from ocean surfaces or from treetops. The sun also provides the energy which drives the weather systems which move the water vapor (clouds) from one place to another (otherwise, it would only rain over the oceans). Precipitation occurs when water condenses from a gaseous state in the atmosphere and falls to earth. Evaporation is the reverse process in which liquid water becomes gaseous. Once water condenses, gravity takes over and the water is pulled to the ground. Gravity continues to operate, either pulling the water underground (groundwater) or across the surface (runoff). In either event, gravity continues to pull water lower and lower until it reaches the oceans (in most cases; the Great Salt Lake, Dead Sea, Caspian Sea, and other such depressions may also serve as the lowest basin into which water can be drawn). Frozen water may be trapped in cooler regions of the Earth (the poles, glaciers on mountaintops, etc.) as snow or ice, and may remain as such for very long periods of time. Lakes, ponds, and wetlands form where water is temporarily trapped. The oceans are salty because any weathering of minerals that occurs as the water runs to the ocean will add to the mineral content of the water, but water cannot leave the oceans except by evaporation, and evaporation leaves the minerals behind. Thus, rainfall and snowfall are comprised of relatively clean water, with the exception of pollutants (such as acids) picked up as the water falls through the atmosphere. Organisms play an important role in the water cycle. As you know, most organisms contain a significant amount of water (up to 90% of their body weight). This water is not held for any length of time and moves out of the organism rather quickly in most cases. Animals and plants lose water through evaporation from the body surfaces, and through evaporation from the gas exchange structures (such as lungs). In plants, water is drawn in at the roots and moves to the gas exchange organs, the leaves, where it evaporates quickly. This special case is called transpiration because it is responsible for so much of the water that enters the atmosphere. In both plants and animals, the breakdown of carbohydrates (sugars) to produce energy (respiration) produces both carbon dioxide and water as waste products. Photosynthesis reverses this reaction, and water and carbon dioxide are combined to form carbohydrates. Now you understand the relevance of the term carbohydrate; it refers to the combination of carbon and water in the sugars we call carbohydrates.

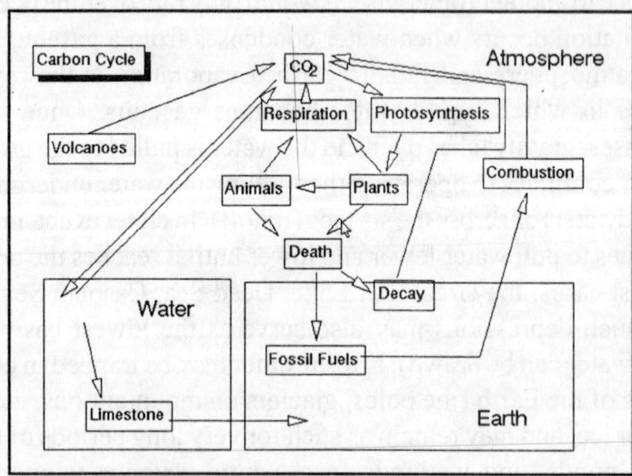

Carbon Cycle

Once you understand the water cycle, the carbon cycle is relatively simple. From a biological perspective, the key events here are the complementary reactions of respiration and photosynthesis. Respiration takes carbohydrates and oxygen and combines them to produce carbon dioxide, water, and energy. Photosynthesis takes carbon dioxide and water and produces carbohydrates and oxygen. The outputs of respiration are the inputs of photosynthesis, and the outputs of photosynthesis are the inputs of respiration. The reactions are also complementary in the way they deal with energy. Photosynthesis takes energy from the sun and stores it in the carbon-carbon bonds of carbohydrates; respiration releases that energy. Both plants and animals carry on respiration, but only plants (and other producers) can carry on photosynthesis. The chief reservoirs for carbon dioxide are in the oceans and in rock. Carbon dioxide dissolves readily in water. Once there, it may precipitate (fall out of solution) as a solid rock known as calcium carbonate (limestone). Corals and algae encourage this reaction and build up limestone reefs in the process. On land and in the water, plants take up carbon dioxide and convert it into carbohydrates through photosynthesis. This carbon in the plants now has 3 possible fates. It can be liberated to the atmosphere by the plant through respiration; it can be eaten by an animal, or it can be present in the plant when the plant dies. Animals obtain all their carbon in their food, and, thus, all carbon in biological systems ultimately comes from plants (autotrophs). In the animal, the carbon also has the same 3 possible fates. Carbon from plants or animals that is released to the atmosphere through respiration will either be taken up by a plant in photosynthesis or dissolved in the oceans. When an animal or a plant dies, 2 things can happen to the carbon in it. It can either be respired by decomposers (and released to the atmosphere), or it can be burried intact and ultimately form coal, oil, or natural gas (fossil fuels). The fossil fuels can be mined and burned in the future; releasing carbon dioxide to the atmosphere. Otherwise, the carbon in limestone or other sediments can only be released to the atmosphere when they are subducted and brought to volcanoes, or when they are pushed to the surface and slowly weathered away.

Humans have a great impact on the carbon cycle because when we burn fossil fuels we release excess carbon dioxide into the atmosphere. This means that more carbon dioxide goes into the oceans, and more is present in the atmosphere. The latter condition causes global warming, because the carbon dioxide in the atmosphere allows more energy to reach the Earth from the sun than it allows to escape from the Earth into space.

The Oxygen Cycle

If you look back at the carbon cycle, you will see that we have also described the oxygen cycle, since these atoms often are combined. Oxygen is present in the carbon dioxide, in the carbohydrates, in water, and as a molecule of two oxygen atoms. Oxygen is released to the atmosphere by autotrophs during photosynthesis and taken up by both autotrophs and heterotrophs during respiration. In fact, all of the oxygen in the atmosphere is *biogenic*; that is, it was released from water through photosynthesis by autotrophs. It took about 2 billion years for autotrophs (mostly cyanobacteria) to raise the oxygen content of the atmosphere to the 21% that it is today; this opened the door for complex organisms such as multicellular animals, which need a lot of oxygen.

The Nitrogen Cycle

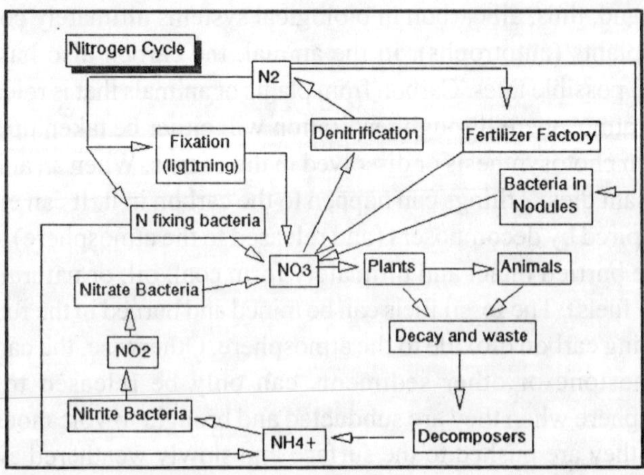

The nitrogen cycle is one of the most difficult of the cycles to learn, simply because there are so many important forms of nitrogen, and because organisms are responsible for each of the interconversions. Remember that nitrogen is critically important in forming the amino portions of the amino acids which in turn form the proteins of your body. Proteins make up skin and muscle, among other important structural portions of your body, and all enzymes are proteins. Since enzymes carry out almost all of the chemical reactions in your body, it's easy to see how important nitrogen is. The chief reservoir of nitrogen is the atmosphere, which is about 78% nitrogen. It is here we reach one of the limits of the hypertext language currently (1995-1996) most in vogue on the WWW. Nitrogen gas in the atmosphere is composed of two nitrogen atoms bound to each other. It is a pretty non-reactive gas; it takes a lot of energy to get nitrogen gas to break up and combine with other things, such as carbon or oxygen. Nitrogen gas can be taken from the atmosphere (fixed) in two basic ways. First, lightning provides enough energy to "burn" the nitrogen and fix it in the form of nitrate, which is a nitrogen with three oxygens attached. This process is duplicated in fertilizer factories to produce nitrogen fertilizers. The other form of nitrogen fixation is by nitrogen fixing bacteria, who use special enzymes instead of the extreme amount of energy found in lightning to fix nitrogen. These nitrogen-fixing bacteria come in three forms: some are free-living in the soil; some form symbiotic, mutualistic associations with the roots of bean plants and other legumes (rhizobial bacteria); and the third form of nitrogen-fixing bacteria are the photosynthetic cyanobacteria (blue-green algae) which are found most commonly in water. All of these fix nitrogen, either in the form of nitrate or in the form of ammonia (nitrogen with 3 hydrogens attached). Most plants can take up nitrate and convert it to amino acids. Animals acquire all of their amino acids when they eat plants (or other animals). When plants or animals die (or release waste) the nitrogen is returned to the soil. The usual form of nitrogen returned to the soil in animal wastes or in the output of the decomposers, is ammonia. Ammonia is rather toxic, but, fortunately there are nitrite bacteria in the soil and in the water which take up ammonia and convert it to nitrite, which is nitrogen with two oxygens. Nitrite is also somewhat toxic, but another type of bacteria, nitrate bacteria, take nitrite and convert it to nitrate, which can be taken up by plants to continue the cycle. We now have a cycle set up in the soil (or water), but what returns nitrogen to the air? It turns out that there are denitrifying bacteria which take the nitrate and combine the nitrogen back into nitrogen gas. The nitrogen cycle has some important practical considerations, as anyone who has ever set up a saltwater fish tank has found out. It takes several weeks to set up such a tank, because you must have sufficient numbers of nitrite and nitrate bacteria present to detoxify the ammonia produced by the fish and decomposers in the tank. Otherwise, the ammonia levels in the tank will build up and kill the fish. This is usually not a problem in freshwater tanks for two reasons. One, the pH in a freshwater tank is at a different level than in a saltwater tank. At the pH of a freshwater tank, ammonia is not as toxic. Second, there are more multicellular plant forms that can grow in freshwater, and these plants remove the ammonia from the water very efficiently. It is hard to get enough plants growing in a saltwater tank to detoxify the water in the same way.

The Phosphorous Cycle

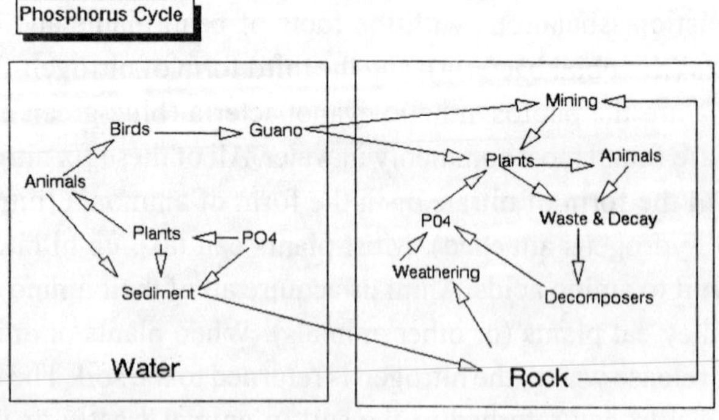

The phosphorous cycle is the simplest of the cycles that we will examine. For our purposes, phosphorous has only one form, phosphate, which is a phosphorous atom with 4 oxygen atoms. This heavy molecule never makes its way into the atmosphere, it is always part of an organism, dissolved in water, or in the form of rock. When rock with phosphate is exposed to water (especially water with a little acid in it), the rock is weathered out and goes into solution. Autotrophs take this phosphorous up and use it in a variety of ways. It is an important constituent of cell membranes, DNA, RNA, and, of course ATP, which, after all, stands for adenosine tri*phosphate*. Heterotrophs (animals) obtain their phosphorous from the plants they eat, although one type of heterotroph, the fungi, excel at taking up phosphorous and may form mutualistic symbiotic relationships with plant roots. These relationships are called *mycorrhizae*; the plant gets phosphate from the fungus and gives the fungus sugars in return. Animals, by the way, may also use phosphorous as a component of bones, teeth and shells. When animals or plants die (or when animals defecate), the phosphate may be returned to the soil or water by the decomposers. There, it can be taken up by another plant and used again. This cycle will occur over and over until at last the phosphorous is lost at the bottom of the deepest parts of the ocean, where it becomes part of the sedimentary rocks forming there. Ultimately, this phosphorous will be released if the rock is brought to the surface and weathered. Two types of animals play a unique role in the phosphorous cycle. Humans often mine rock rich in phosphorous. For instance, in Florida, which was once sea floor, there are extensive phosphate mines. The phosphate is then used as fertilizer. This mining of phosphate and use of the phosphate as fertilizer greatly accelerates the phosphorous cycle and may cause local overabundance of phosphorous, particularly in coastal regions, at the mouths of rivers, and anyplace where there is a lot of sewage released into the water (the phosphate placed on crops finds its way into our stomachs and from there to our toilets). Local abundance of phosphate can cause overgrowth of algae in the water; the algae can use up all the oxygen in the water and kill other aquatic life. This is called eutrophication. The other animals that play a unique role in the phosphorous cycle are marine birds. These birds take phosphorous containing fish out of the ocean and return to land, where they defecate. Their guano contains high levels of phosphorous and in this way marine birds return phosphorous from the ocean to the land. The guano is often mined and may form the basis of the economy in some areas.

Forest ecosystem

An **ecosystem** is a natural unit consisting of all plants, animals and micro-organisms (biotic factors) in an area functioning together with all of the non-living physical (abiotic) factors of the environment.

Scope of forest ecology

Forest ecology is one branch of a biotically-oriented classification of types of ecological study (as opposed to a classification based on organizational level or complexity, for example population or community ecology). Thus, forests can be, and are, studied at any number of organizational levels, from the individual organism to the ecosystem. However, as the term forest connotes an area inhabited by more than one organism, forest ecology most often concentrates on the level of the population, community or ecosystem. Logically, trees are an important component of forest research, but the wide variety of other life forms and abiotic components in most forests means that other elements, such as wildlife or soil nutrients, are often the focal point. Thus, forest ecology is a highly diverse and important branch of ecological study.

Grassland ecosystem

Grasslands (also called **greenswards**) are areas where the vegetation is dominated by grasses (Poaceae) and other herbaceous (non-woody) plants (forbs). However, sedge (Cyperaceae) and rush (Juncaceae) families can also be found. Grasslands occur naturally on all continents except Antarctica. In temperate latitudes, such as northwest Europe, grasslands are dominated by perennial species, whereas in warmer climates annual species form a greater component of the vegetation.

Grasslands can be found in most terrestrial climates. Grassland vegetation can vary in height from very short, as in chalk downland where the vegetation may be less than 30 cm (12 in) high, to quite tall, as in the case of North American tallgrass prairie, South American grasslands and African savanna. Woody plants, shrubs or trees, may occur on some grasslands - forming savannas, scrubby grassland or semi-wooded grassland, such as the African savannas. Such grasslands are sometimes referred to as wood-pasture or woodland.

Grasslands cover nearly fifty percent of the land surface of the continent of Africa. While grasslands in general support diverse wildlife, given the lack of hiding places for predators, the African Savanna regions support a much greater diversity in wildlife than do temperate

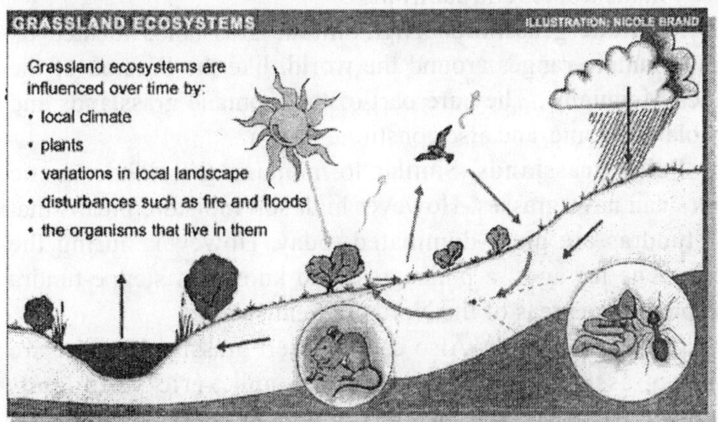

Types of grassland

Tropical and subtropical grasslands–These grasslands are classified with tropical and subtropical savannas and shrublands as the tropical and subtropical grasslands, savannas, and shrublands biome. Notable tropical and subtropical grasslands include the Llanos grasslands of northern South America.

Temperate grasslands–Mid-latitude grasslands, including the Prairie of North America, the Pampa of Argentina, calcareous downland, and the steppes of Europe. They are classified with temperate savannas and shrublands as the temperate grasslands, savannas, and shrublands biome. Temperate grasslands are the home to many large herbivores, such as bison, gazelles, zebras, rhinoceroses, and wild horses. Carnivores like lions, wolves and cheetahs and leopards are also found in temperate grasslands. Other animals of this region include: deer, prairie dogs, mice, jack rabbits, skunks, coyotes, snakes, fox, owls, badgers, blackbirds, grasshoppers, meadowlarks, sparrows, quails, hawks and hyenas.

Flooded grasslands–Grasslands that are flooded seasonally or year-round, like the Everglades of Florida or the Pantanal of Brazil, Bolivia and Paraguay.They are classified with flooded savannas as the flooded grasslands and savannas biome and occur mostly in the tropics and subtropics.

Montane grasslands–High-altitude grasslands located on high mountain ranges around the world, like the Páramo of the Andes Mountains. They are part of the montane grasslands and shrublands biome, and also constitute tundra.

Polar grasslands–Similar to montane grasslands, arctic tundra can have grasses. However high soil moisture means that few tundras are grass-dominated today. However, during the Pleistocene ice ages, a polar grassland known as steppe-tundra occupied large areas of the Northern hemisphere.

Xeric grasslands–Also called desert grasslands, these are sparse grasslands located in deserts and xeric shrublands ecoregions.

Estuaries

An **estuary** is a semi-enclosed coastal body of water with one or more rivers or streams flowing into it, and with a free connection to the open sea. Estuaries are often associated with high levels of biological diversity.

Estuaries are typically the tidal mouths of rivers (*aestus* is Latin for tide), and they are often characterized by sedimentation or silt carried in from terrestrial runoff and, frequently, from offshore. They are made up of brackish water. Estuaries are often given names like bay, sound, etc. The terms are not mutually exclusive.

As ecosystems, many estuaries are under threat from human activities such as pollution and overfishing. Estuaries are more likely to occur on submerged coasts, where the sea level has risen in relation to the land; this process floods valleys to form rias. These can become estuaries if there is a stream or river flowing into them.

Types of estuary

Estuaries can be grouped by circulation.

- **Salt wedge**. River output greatly exceeds marine input; there is little mixing, and thus a sharp contrast between fresh surface water and saline bottom water.
- **Highly stratified**. River output and marine input are more even, with river flow still dominant; turbulence induces more mixing of salt water upward than the reverse.
- **Slightly stratified**. River output is less than the marine input. Here, turbulence causes mixing of the whole water column, such that salinity varies more longitudinally rather than vertically.
- **Vertically mixed**. River output is much less than marine input, such that the freshwater contribution is negligible; longitudinal salinity variation only.
- **Inverse estuary**. Located in regions with high evaporation, there is no freshwater input and in fact salinity increases inland; TNR *inward* at the surface, downwells at the inland terminus, and flows outward subsurface.
- **Intermittent estuary**. Estuary type varies dramatically depending on freshwater input, and is capable of changing from a wholly marine embayment to any of the other estuary types.

Aquatic ecosystem

An **aquatic ecosystem** is an ecosystem located in water bodies. Communities of organisms that are dependent on each other and on their environment live in aquatic ecosystems. The two main types of aquatic ecosystems are marine ecosystems and freshwater ecosystems.

Marine ecosystems are among of the earth's aquatic ecosystems. They include oceans, salt marsh and intertidal ecology, estuaries and lagoons, mangroves and coral reefs, the deep sea and the sea floor. They can be contrasted with freshwater ecosystems, which have a smaller salt content.

Such places are considered ecosystems because the plant life supports the animal life and vice-versa.Marine ecosystems cover approximately 71% of the Earth's surface and contain approximately 97% of the planet's water. They generate 32% of the world's net primary production. They are distinguished from freshwater ecosystems by the presence of dissolved compounds, especially salts, in the water. Approximately 85% of the dissolved materials in seawater are sodium and chlorine. Seawater has an average salinity of 35 parts per thousand (ppt) of water. Actual salinity varies among different marine ecosystems.

Freshwater ecosystems are among the earth's aquatic ecosystems. They include lakes and ponds, rivers, streams and springs, and wetlands. They can be contrasted with marine ecosystems, which have a larger salt content. Freshwater habitats can be classified by different factors, including temperature, light penetration, and vegetation.

Freshwater ecosystems cover 0.8% of the Earth's surface and contain 0.009% of its total water. They generate nearly 3% of its net primary production. Freshwater ecosystems contain 41% of the world's known fish species.

There are three basic types of freshwater ecosystems:

- Lentic: slow-moving water, including pools, ponds, and lakes.
- Lotic: rapidly-moving water, for example streams and rivers.
- Wetlands: areas where the soil is saturated or inundated for at least part of the time.

3

NATURAL RESOURCES

The Earth's natural resources are vital to the survival and development of the human population. Some of these resources, such as minerals, species, and habitats, are finite — once they have been exhausted or destroyed, they are gone forever. Others, such as air, water, and wood, are renewable — although we generally rely on the Earth's natural systems to regrow, renew, and purify them for us. Although many effects of over-exploitation are felt locally, the growing interdependence of nations and international trade in natural resources make their management a global issue. Meaning of natural resources:

1. Substance formed naturally that is known to be useful.
2. Essential for heating, lighting and driving machinery.
3. Cities, industries and transport networks cannot function without it.

Natural resources (economically referred to as **land** or **raw materials**) are naturally forming substances that are considered valuable in their relatively unmodified (natural) form. A natural resource's value rests in the amount and extractability of the material available and the demand for it. The latter is determined by its usefulness to production. A commodity is generally considered a natural resource when the primary activities associated with it are extraction and purification, as opposed to creation. Thus, mining, petroleum extraction, fishing, hunting, and forestry are generally considered natural-resource industries, while agriculture is not. The Nation's natural resources include its minerals, energy, land, water, and biota. The earth's natural resources are finite, which means that if we use them continuously, we will eventually exhaust them.

A **Natural resource** is a thing people can use which comes from Nature: people do not make it themselves. Examples of natural resources are air, water, wood, oil, solar energy, wind energy, coal water, minerals. Petroleum used in cars is not a natural resource, for example, because people make it.

We often say there are two sorts of natural resource: *renewable resources* and *non-renewable resources*.

- A renewable resource grows again or comes back again after we used it. For example, fish.
- A non-renewable resource is a resource that does not grow or come back, or a resource that would take a very long time to come back or grow. For example, coal is a non-renewable resource. When we use coal, there is less coal afterwards. One day, there will be no more a lot of natural resources to make goods. They can use a resource directly (for example, eating the fish or burning the wood to cook the fish), or they can change it by industry into a different thing (for example, they can use wind energy to make electricity to cook the fish).

In every country or place can get all natural resources. When people do not have one resource where they live, they can a) use another resource, or b) trade with another country (for example, they can buy oil from their neighbours). Some resources are rare - difficult to find–so people sometimes fight to have them (for example, oil resources).

When people do not have some natural resources their quality of life can drop. For example, when they can not get clean water, people may become ill; if there is not enough wood, all trees will be cut and the forest will disappear (deforestation); if there are not enough fish in a sea, people can die of starvation. Some examples of renewable resources are wood, solar energy, trees, wind, hydroelectric power, fish and sunlight.

Classification of Natural Resources

Natural resources are mostly classified into **renewable** and **non-renewable resources**. Sometimes resources are classified as non-renewable even if they are technically renewable, just not easily renewed within a reasonable amount of time, such as fossil fuels.

Renewable Resources

A Natural Resources qualifies as a **renewable resource** if it is replenished by natural processes at a rate comparable or faster than its rate of consumption by humans or other users. Solar radiation, tides, winds and hydroelectricity are *perpetual resources* that are in no danger of long-term availability. Renewable resources may also mean commodities such as wood, paper, and leather.

Some natural renewable resources such as geothermal power, fresh water, timber, and biomass must be carefully managed to avoid exceeding the environment's capacity to replenish them. A life cycle assessment provides a systematic means of evaluating renewability.

Renewable resources are sometimes living resources (trees and soil, for example), which can restock (renew) themselves if they are not over-harvested and used sustainably. There are also non-living resources that are renewable, such as hydroelectric power, solar power, biomass fuel, and wind power. If renewable resources are consumed at a rate above their natural rate of replacement, the standing stock will diminish and eventually run out. The rate of sustainable use of a renewable resource is determined by the replacement rate and amount of standing stock of that particular resource. Non-living renewable natural resources include dirt and water.

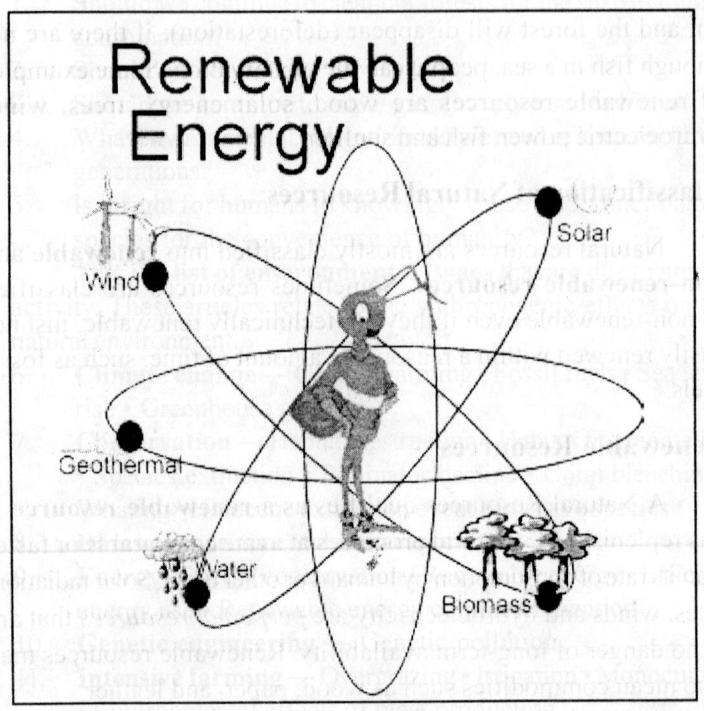

Flow renewable resources are very much like renewable resources, only they do not need regeneration, unlike renewable resources. Flow renewable resources include renewable energy sources such as the following renewable power sources: solar, geothermal, landfill gas, tides and wind.

Resources can also be classified on the basis of their origin as **biotic** and **abiotic**. Biotic resources are derived from living organisms. Abiotic resources are derived from the non-living world (e.g., land, water, and air). Mineral and power resources can be abiotic natural resources.

Renewable energy is energy generated from natural resources—such as sunlight, wind, rain, tides and geothermal heat—which are renewable (naturally replenished). In 2006, about 18% of global final energy consumption came from renewables, with 13% coming from traditional biomass, such as wood-burning.

Hydroelectricity was the next largest renewable source, providing 3% (15% of global electricity generation), followed by solar hot water/heating, which contributed 1.3%. Modern technologies, such as geothermal energy, wind power, solar power, and ocean energy together provided some 0.8% of final energy consumption.

Types of Renewable Energy Sources

The renewable sources of energy mainly include solar, wind, hydel, biomass, geothermal and ocean energies.

1. Solar energy or the energy from the sun is the most sought after. It is a non-polluting and constant source of energy. Equipment designed for this purpose can be used for many functions, and hence it is also cost effective. Equipment to harness solar energy can be built both for domestic and industrial purposes. Though availability of solar energy depends on the availability of sunlight at all times, it is highly beneficial in the sense that it causes no pollution at all.

2. Wind energy is another good source of renewable energy. This is particularly feasible in regions like coastlines and high altitude areas which receive strong wind currents throughout the year. Energy produced from a single windmill can be used for domestic purposes. A number of windmills coupled together help to generate energy on a large scale.

3. Water energy is available in a number of forms. For instance, the water which is stored in dams can be released from a great height and made to flow through turbines, hence converting potential energy of falling water into kinetic energy. The oceans also possess a high amount of energy, either as wave energy present in the waves or as energy which can be generated due to difference in temperatures of different regions of the ocean.

4. Biomass, which is nothing but human and animal bi-products, can be used to generate energy. This is a very useful method, as it helps to cleanse the environment of the unwanted waste matter and also provides a non-polluting source of energy. This method has come into limelight due to problem of waste disposal

being faced in many of the major towns and cities. Apart from this, there is also the geothermal energy which can be obtained from heat energy present either in the earth's layers or on the earth's surface.

As you can see, there are a large number of renewable energy sources which can be accessed to the benefit of the human race.

Non-Renewable Resource is a natural resource that cannot be produced, re-grown, regenerated, or reused on a scale which can sustain its consumption rate. These resources often exist in a fixed amount, or are consumed much faster than nature can recreate them. Fossil fuels (such as coal, petroleum and natural gas) and nuclear fuel are some examples. In contrast, resources such as timber (when harvested sustainably) or metals (which can be recycled) are considered renewable resources.A non-renewable resource is always drawn down with anabolic processes that use up energy.

Natural resources such as coal, petroleum, and natural gas take millions of years to form naturally and cannot be replaced as fast as they are being consumed. Eventually natural resources will become too costly to harvest and humanity will need to find other sources of energy. At present, the main energy sources used by humans are non-renewable.

Some natural resources, called renewable resources, are replaced by natural processes given a reasonable amount of time. Soil, water, forests, plants, and animals are all renewable resources as long as they are properly conserved. Solar, wind, wave, and geothermal energies are based on renewable resources. Renewable resources such as the movement of water (hydropower, including tidal power; ocean surface waves used for wave power), wind (used for wind power), geothermal heat (used for geothermal power); and radiant energy (used for solar power) are practically infinite and cannot be depleted, unlike their non-renewable counterparts, which are likely to run out if not used wisely. Still, these technologies are not fully utilized.

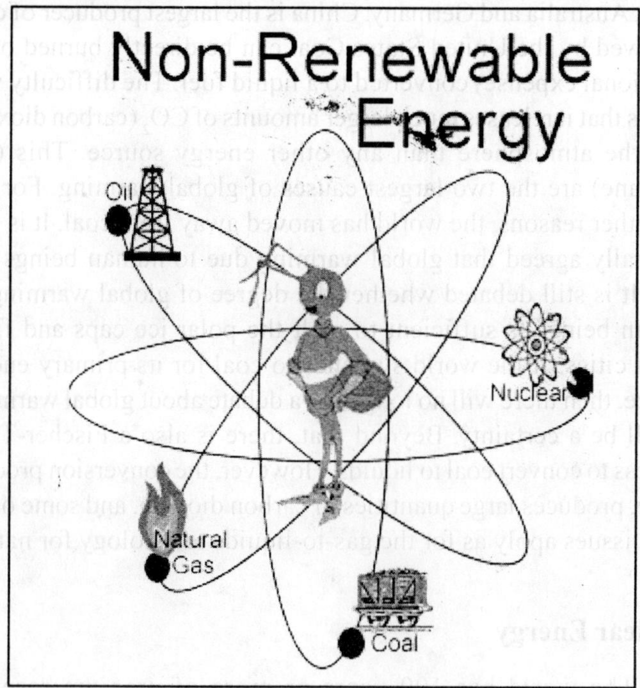

Natural Gas

Natural gas is also a finite resource, but its production plateau for the world may be ten years off or further. 54% of the world's proven reserves is in Russia, Iran and Qatar.North America has reached its plateau now, and will soon be a major importer. Natural gas can and is imported from other continents by first lowering the temperature to -259 Celsius) to form liquefied natural gas (LNG). There are (expensive) schemes to use natural gas as fuel instead of gasoline. An alternative to natural gas as fuel is to use fuel cells or batteries to convert other stationary energy production into mobile energy production.

Coal

Coal reserves are estimated to last 100 years or more .78% of the world's proved reserves of coal is in the U.S., Russia, China

India, Australia and Germany. China is the largest producer of coal, followed by the United States.Coal can be directly burned or (at additional expense) converted to a liquid fuel. The difficulty with coal is that it releases much larger amounts of CO_2 (carbon dioxide) into the atmosphere than any other energy source. This (and methane) are the two largest causes of global warming. For this and other reasons, the world has moved away from coal. It is now generally agreed that global warming due to human beings is a fact. It is still debated whether the degree of global warming by human beings is sufficient to melt the polar ice caps and flood many cities. If the world switches to coal for its primary energy source, then there will no longer be a debate about global warming. It will be a certainty. Beyond that, there is also a Fischer-Tropf process to convert coal to liquids. However, the conversion process, alone, produces large quantities of carbon dioxide, and some of the same issues apply as for the gas-to-liquids technology for natural gas.

Nuclear Energy

The world has 100 years or more of uranium reserves (especially if the world also uses breeder reactors to reprocess expended fuel).The United States, France, Japan, Germany, Russia and South Korea produce 74% of the world's nuclear energy, as electricity .There are also worries about nuclear weapon proliferation. Nuclear reactors produce nuclear waste with trace amounts of plutonium and fissionable uranium. If every country has nuclear reactors, then every country has the potential to reprocess the expended nuclear fuel into weapons-grade uranium or plutonium. It also opens the possibility for smaller groups to steal such expended fuel for reprocessing elsewhere. If countries use breeder reactors to reprocess expended fuel, then it is even easier to convert a portion into weapons-grade uranium or plutonium.

Fusion

Fusion is an energy source that has always appeared to be twenty years away from commercialization. Fusion reactors consume deuterium (which can be extracted in abundance from water) and lithium (which can be abundantly mined). The world is planning a $5 billion demonstration International Thermonuclear Experimental Reactor (ITER) project. Current potential participants include the European Union, Japan, Russia, United States, China and Korea. Eight years after the beginning of construction, it will be commissioned and start producing regular power. If successful a commercial prototype will follow. If that is successful, further commercial fusion reactors will be built. The ITER project is currently delayed by discussions whether it should be sited in Japan or in France.

Shale Oil

Shale oil exists in abundant quantities in the United States and elsewhere. If all of it could be extracted, the United States alone would have more reserves than exist in Saudi Arabia. However, only a fraction of the shale oil has rich enough oil yields to consider for exploitation.

Shale oil is actually kerogen (pre-oil) locked into rocks. Kerogen can be converted into a petroleum-like substance. Most of the world's shale oil (more than 95%) is in the western United States. The difficulty is to extract shale oil in an energetically efficient manner. If more energy is expended in extracting the shale oil than is contained in the shale oil, then the process becomes uneconomic. There are also environmental issues. Shale oil remains today a difficult technological problem, and mass production is not on the near-term horizon.

Tar Sands and Extra-Heavy Oil

Related to shale oil are tar sands. Tar sands (natural bitumen) are a mixture of sand and viscous oil that can be further processed into crude oil. Tar sands and their cousins, extra-heavy oil, are a degraded form of oil. Currently, Canada is producing 1 million barrels per day of oil from tar sands and heavy oil, with an increase to over 2 million barrels per day planned by 2010. Venezuela has lower goals.

Limitations of Non-renewable Resources

Non-renewable resources consist of geochemical concentrations of naturally occurring elements and compounds that can or may be exploited at a profit. Almost 70 percent of proved crude-oil reserves are n the Middle East, and five countries produce more than 65 percent of the world's copper. These non-uniform distributions are real and do not merely reflect differences in exploration effort. Many of the most common ore deposits appear genetically and spatially related to the boundaries of crustal plates. Other ore deposits not associated with crustal plate margins may be localized by uplift of the crust over thermal plumes. Petroleum as well as metallogenic provinces exist, and the world's coal deposits are strikingly concentrated in the temperate belt of the Northern Hemisphere.

Not only are, geologic resources distributed unevenly over the surface of the globe, they are concentrated in the outermost part of the earth's continental crust. Mechanisms that concentrate chemical elements operate most effectively on and near the earth's surface: weathering, erosion, sorting during transport, groundwater leaching, and supergene enrichment are effective only in the upper few hundred meters of the continents. Hot metal-bearing solutions encounter the precipitating effect of colder meteoric waters and find relatively high permeability only in the upper few thousand meters of the crust. Oil and gas formation, migration, and entrapment take place within the thin and discontinuous sedimentary skin of the continents. The mineral fuel analog is the transition zone in a reservoir rock between crude oil and water, a zone commonly only a meter or two thick; the world's greatest oil well, after producing more than 50 million barrels of oil, suddenly produced only salt water.

The limited nature of individual deposits, the difficulty of seeing through rock, and the history of many mining districts and oil fields have led to many forecasts of depletion and exhaustion. But, as old deposits have become exhausted and new ones have become harder to find, world mineral and fossil-fuel production paradoxically has continued to increase. There is a wide difference of opinion now on the question of limits to the exploitation of non-renewable resources.

Three Kinds of Limits

To the mining or petroleum engineer, the profit of exploitation is defined by the difference between the price received for his product and the pecuniary costs of recovering the natural material and turning it into a saleble product. To society, however, the profit from mining (including oil and gas extraction) can be defined either as an energy surplus, as from the exploitation of fossil and nuclear fuel deposits, or as a work saving, as in the lessened expenditure of human energy and time when steel is used in place of wood in tools and structures. In this context, the exploitation of earth resources for display, adornment, or monetary backing is a deficit operation, financed by energy profits from other kinds of mining.

The ultimate limit to exploitation of earth resources then is the limit of net energy profit (or work savings). When it takes more energy to find, recover, process, and transport the fossil fuels than can be gotten from them in useful form, there will be no more oil, gas, or coal resources. When it takes so much energy or work to produce a non-energy material that one must sacrifice other, more needed, items or services to pay for it, there will be no more resources of that material. Short of this energetic limit, one or both of two other limits may intervene. First is the limit of comparable utility. A resource is a resource only while it can be used to perform a function desired by man better or more cheaply than can another substance. If the energy cost of an earth resource such as crude oil rises to a level at which another substance, such as synthetic crude oil from coal, can be substituted in adequate volume at a lower cost for comparable utility, substitution will take place, with the first resource reverting to a mere geochemical anomaly and the source of the replacing substance becoming a resource.

The limit to a resource also may be determined by the unwillingness of society to pay the cost of its exploitation, even though an energy surplus (or saving) might be obtained thereby. A modern decision to for go the calculable energy benefits of the breeder-reactor power plant would again invoke this limit of living-level degradation, a limit that comes into play when a society is not willing to pay the total costs of production-because so doing would, it is judged, lower the level of living more than would forgoing use of the resource.

Natural Resources And Sustainable Development

Natural resources are naturally occurring substances that are considered valuable in their relatively unmodified (natural) form. A natural resource's value rests in the amount and extractability of the material available and the demand for it. The latter is determined by its usefulness to production. A commodity is generally considered a natural resource when the primary activities associated with it are extraction and purification, as opposed to creation. Thus, mining, petroleum extraction, fishing, hunting, and forestry are generally considered natural-resource industries, while agriculture is not.

A **Natural resource** is a thing people can use which comes from Nature: people do not make it themselves. Examples of natural resources are air, water, wood, oil, solar energy, wind energy, coal water, minerals. Petroleum used in cars is not a natural resource, for example, because people make it.

We often say there are two sorts of natural resource: *renewable resources* and *non-renewable resources*.

- A renewable resource grows again or comes back again after we used it. For example, fish.
- A non-renewable resource is a resource that does not grow or come back, or a resource that would take a very long time to come back or grow. For example, coal is a non-renewable resource. When we use coal, there is less coal afterwards. One day, there will be no more a lot of natural resources to make goods. They can use a resource directly (for example, eating the fish or burning the wood to cook the fish), or they can change it by industry into a different thing (for example, they can use wind energy to make electricity to cook the fish).

In every country or place can get all natural resources. When people do not have one resource where they live, they can : a) use another resource, or b) trade with another country (for example, they can buy oil from their neighbours). Some resources are rare - difficult to find - so people sometimes fight to have them (for example, oil resources).

When people do not have some natural resources their quality of life can drop. For example, when they can not get clean water, people may become ill; if there is not enough wood, all trees will be cut and the forest will disappear (deforestation); if there are not enough fish in a sea, people can die of starvation. Some examples of renewable resources are wood, solar energy, trees, wind, hydroelectric power, fish and sunlight.

Sustainable development is a pattern of resource use that aims to meet human needs while preserving the environment so that these needs can be met not only in the present, but in the indefinite future. Sustainable development ties together concern for the carrying capacity of natural systems with the social challenges facing humanity. Ecologists have pointed to the "limits of growth" and presented the alternative of a "steady state economy" in order to address environmental concerns.

The field of sustainable development can be conceptually broken into three constituent parts: environmental sustainability, economic sustainability and sociopolitical sustainability.

Environmental sustainability

Environmental sustainability is the process of making sure current processes of interaction with the environment are pursued with the idea of keeping the environment as pristine as naturally possible based on ideal-seeking behaviour.

An "unsustainable situation" occurs when natural capital (the sum total of nature's resources) is used up faster than it can be replenished. Sustainability requires that human activity only uses nature's resources at a rate at which they can be replenished naturally. Inherently the concept of sustainable development is intertwined with the concept of carrying capacity. Theoretically, the long-term result of environmental degradation is the inability to sustain human life. Such degradation on a global scale could imply extinction for humanity.

Economic Sustainability

Economic Sustainability identified information, integration, and participation as key building blocks to help countries achieve development that recognises these interdependent pillars. It emphasises that in sustainable development everyone is a user and provider of information. It stresses the need to change from old sector-centred ways of doing business to new approaches that involve cross-sectoral co-ordination and the integration of environmental and social concerns into all development processes.

Sustainability is a process which tells of a development of all aspects of human life affecting sustenance. It means resolving the conflict between the various competing goals, and involves the simultaneous pursuit of economic prosperity, environmental quality and social equity famously known as three dimensions (triple bottom line) with is the resultant vector being technology, hence it is a continually evolving process; the 'journey' (the process of achieving sustainability) is of course vitally important, but only as a means of getting to the destination (the desired future state). However, the 'destination' of sustainability is not a fixed place in the normal sense that we understand destination. Instead, it is a set of wishful characteristics of a future system.

4

FOREST RESOURCES

A **forest** is an area with a high density of trees. There are many definitions of a forest, based on various criteria.These plant communities cover approximately 9.4% of the Earth's surface (or 30% of total land area) and function as habitats for organisms, hydrologic flow modulators, and soil conservers, constituting one of the most important aspects of the Earth's biosphere. Historically, "forest" meant an uncultivated area legally set aside for hunting by feudal nobility, and these hunting forests were not necessarily wooded much if at all. However, as hunting forests did often include considerable areas of woodland, the word forest eventually came to mean wooded land more generally. A woodland is ecologically distinct from a forest.

Forests can be found in all regions capable of sustaining tree growth, at altitudes up to the tree line, except where natural fire frequency or other disturbance is too high, or where the environment has been altered by human activity. As a general rule, forests dominated by angiosperms (*broadleaf forests*) are more species-rich than those dominated by gymnosperms (*conifer*, *montane*, or *needleleaf forests*), although exceptions exist. Forests sometimes contain many tree species within a small area (as in tropical rain and temperate deciduous forests), or relatively few species over large areas (e.g., taiga and arid montane coniferous forests). Forests are often home to many animal and plant species, and biomass per unit area is high compared to other vegetation communities. Much of this biomass occurs below ground in the root systems and as partially decomposed plant detritus. The woody component of a forest contains lignin, which is relatively slow to decompose compared with other organic materials such as cellulose or carbohydrate.

Forests are differentiated from woodlands by the extent of canopy coverage: in a forest the branches and the foliage of separate trees often meet or interlock, although there can be gaps of varying sizes within an area referred to as forest. A woodland has a more continuously open canopy, with trees spaced further apart, which allows more sunlight to penetrate to the ground between them.

The scientific study of forest species and their interaction with the environment is referred to as forest ecology, while the management of forests is often referred to as forestry.

More forest fixes CO2 emissions

Through many years less trees have been taken out from Norwegian forests than the annual increment. This means that the total biomass of forest is increasing. When the biomass increases, more carbondioxide is fixed in the trees. The greenhouse effect "is stored" in standing forest.

Calculations indicate that the biomass of standing trees has almost doubled since 1925. There are several causes to this increase. One is the planting of forest on areas that previously were used for farmland. Another reason is fertilising, especially the indirect fertilising in the form of nitrous compounds deposited by acid rain. Clear-cutting and introduction of new species of trees have also led to a more rapid growth.

The increased biomass of trees has partly been on the expense of the biological diversity. Much of the present forest growth is done through large monocultures. Large stands of the same year class will also contribute to a reduction in the number of species of flora and fauna that is otherwise found in a more mixed type of vegetation.The planting of trees also reduces the arable land for food production.

Changes in standing volume

The volume of forest increases as a result of growth of the trees and afforestation, or planting, and decreases as a result of felling and natural loss. The increase can be attributed to many factors; intensive forest management, afforestation, return of natural vegetation to uncultivated land and to fertilisation. This includes fertilisation from long-range transport of nitrogen in precipitation.

Calculations of the cubic mass show that the volume of standing forest increased by more than 95 per cent from 1925 to 1995. The calculated forest balance for 1995 shows a standing cubic mass of forest of 630 million m^3 This volume was distributed between 46 per cent spruce, 33 per cent pine and 22 per cent deciduous trees. In 1995 the net increase (increment minus removal) in cubic mass was 9.8 million m^3, or 1.6 per cent of the total cubic mass of standing forest. The net increase was greatest for deciduous trees and pine. It should be noted that the steep increment the last years is partly due to new methods of calculation.

Although the volume of forest has increased considerably since the turn of the century, today's type of forest is different. Clear-cutting, afforestation, introduction of alien species, ditching, building of forest roads, acidification and pollution are among the factors that are affecting the forest as a natural resource and the biological diversity in the forests.

TYPES OF FOREST IN INDIA

India has a diverse range of forests: from the rainforest of Kerala in the south to the alpine pastures of Ladakh in the north, from the deserts of Rajasthan in the west to the evergreen forests in the north-east. Climate, soil type, topography, and elevation are the main factors that determine the type of forest. Forests are classified according to their nature and composition, the type of climate in which they thrive, and its relationship with the surrounding environment.

Forests can be divided into six broad types, with a number of sub types.

Moist tropical	**Montane sub tropical**
Wet evergreen	Broad leaved
Semi-evergreen	Pine
Moist deciduous	Dry evergreen
Littoral and swamp	
Dry tropical	**Montane temperate forests**
Dry deciduous	Wet
Thorn	Moist
Dry evergreen	Dry
Sub alpine	**Alpin**
	Moist
	Dry

Moist tropical forests

Wet evergreen : Wet evergreen forests are found in the south along the Western Ghats and the Nicobar and Andaman Islands and all along the north-eastern region. It is characterized by tall, straight evergreen trees that have a buttressed trunk or root on three sides like a tripod that helps to keep a tree upright during a storm. These trees often rise to a great height before they open out like a cauliflower. The more common trees that are found here are the jackfruit, betel nut palm, jamun, mango, and hollock. The trees in this forest form a tier pattern: shrubs cover the layer closer to the ground, followed by the short structured trees and then the tall variety. Beautiful fern of various colours and different varieties of orchids grow on the trunks of the trees.

Semi-evergreen : Semi-evergreen forests are found in the Western Ghats, Andaman and Nicobar Islands, and the Eastern Himalayas. Such forests have a mixture of the wet evergreen trees and the moist deciduous tress. The forest is dense and is filled with a large variety of trees of both types.

Moist deciduous : Moist deciduous forests are found throughout India except in the western and the north-western regions. The trees have broad trunks, are tall and have branching trunks and roots to hold them firmly to the ground. Some of the taller trees shed their leaves in the dry season. There is a layer of shorter trees and evergreen shrubs in the undergrowth. These forests are dominated by sal and teak, along with mango, bamboo, and rosewood.

Littoral and swamp: Littoral and swamp forests are found along the Andaman and Nicobar Islands and the delta area of the Ganga and the Brahmaputra. It consists mainly of whistling pines, mangrove dates, palms, and bulletwood. They have roots that consist of soft tissue so that the plant can breathe in the water.

Dry tropical forests

Dry deciduous forest : Dry deciduous forests are found throughout the northern part of the country except in the North-East. It is also found in Madhya Pradesh, Gujarat, Andhra Pradesh, Karnataka, and Tamil Nadu. The canopy of the trees does not normally exceed 25 metres. The common trees are the sal, a variety of acacia, and bamboo.

Thorn : This type is found in areas with black soil: North, West, Central, and South India. The trees do not grow beyond 10 metres. Spurge, caper, and cactus are typical of this region.

Dry evergreen : Dry evergreens are found along the Andhra Pradesh and Karnataka coast. It has mainly hard-leaved evergreen trees with fragrant flowers, along with a few deciduous trees.

Montane sub tropical forests

Broad-leaved forests : Broad-leaved forests are found in the Eastern Himalayas and the Western Ghats, along the Silent Valley. There is a marked difference in the form of the vegetation in the two areas. In the Silent Valley, the poonspar, cinnamon, rhododendron, and fragrant grass are predominant. In the Eastern Himalayas, the flora has been badly affected by the shifting cultivation and forest fires. These wet forests consist mainly of evergreen trees with a sprinkling of deciduous here and there. There are oak, alder, chestnut, birch, and cherry trees. There are a large variety of orchids, bamboo and creepers.

Pine : Pine forests are found in the steep dry slopes of the Shivalik Hills, Western and Central Himalayas, Khasi, Naga, and Manipur Hills. The trees predominantly found in these areas are the chir, oak, rhododendron, and pine. In the lower regions sal, sandan, amla, and laburnum are found.

Dry evergreen : Dry evergreen forests normally have a prolonged hot and dry season and a cold winter. It generally has evergreen trees with shining leaves that have a varnished look. Some of the more common ones are the pomegranate, olive, and oleander. These forests are found in the Shivalik Hills and foothills of the Himalayas up to a height of 1000 metres.

Montane temperate forests

Wet : Wet montane temperate forests occur in the North and the South. In the North, it is found in the region to the east of Nepal into Arunachal Pradesh, at a height of 1800–3000 metres, receiving a minimum rainfall of 2000 mm. In the South, it is found in parts of the Niligiri Hills, the higher reaches of Kerala. The forests in the northern region are denser than in the South. This is because over time the original trees have been replaced by fast-growing varieties such as the eucalyptus. Rhododendrons and a variety of ground flora can be found here.

In the North, there are three layers of forests: the higher layer has mainly coniferous, the middle layer has deciduous trees such as the oak and the lowest layer is covered by rhododendron and champa.

Moist : This type spreads from the Western Himalayas to the Eastern Himalayas. The trees found in the western section are broad-leaved oak, brown oak, walnut, rhododendron, etc. In the Eastern Himalayas, the rainfall is much heavier and therefore the vegetation is also more lush and dense. There are a large variety of broad-leaved trees, ferns, and bamboo. Coniferous trees are also found here, some of the varieties being different from the ones found in the South.

Dry : This type is found mainly in Lahul, Kinnaur, Sikkim, and other parts of the Himalayas. There are predominantly coniferous trees that are not too tall, along with broad-leaved trees such as the oak, maple, and ash. At higher elevation, fir, juniper, deodar, and chilgoza can be found.

Sub alpine *:* Sub alpine forests extends from Kashmir to Arunachal Pradesh between 2900 to 3500 metres. In the Western Himalayas, the vegetation consists mainly of juniper, rhododendron, willow, and black currant. In the eastern parts, red fir, black juniper, birch, and larch are the common trees. Due to heavy rainfall and high humidity the timberline in this part is higher than that in the West. Rhododendron of many species covers the hills in these parts.

Alpine

Moist : Moist alpines are found all along the Himalayas and on the higher hills near the Myanmar border. It has a low scrub, dense evergreen forest, consisting mainly of rhododendron and birch. Mosses and ferns cover the ground in patches. This region receives heavy snowfall.

Dry : Dry alpines are found from about 3000 metres to about 4900 metres. Dwarf plants predominate, mainly the black juniper, the drooping juniper, honeysuckle, and willow.

Importance of Forest for Human Society

Providing us with shade and beauty on a largely agricultural and urban landscape.A great emphasis is placed on trees in this Web module. For each forest type, for example, there is a tree list with accompanying descriptions and herbarium sheets. Panoramas of forest types feature-hotspots that link to individual tree descriptions.

This emphasis on trees reflects their ecological, biological, and cultural importance. Also, trees are critical to the classification of forests. Trees represent some of the oldest living organisms on the planet. For example, bristlecone pines, sequoias, and cypress are all long-lived species of trees. Trees were instrumental in the development and support of civilizations. They form important links in the earth's geological, chemical, and hydrological cycles by:

- Taking in CO2 and releasing oxygen;
- Releasing carbon and mineral elements such as nitrogen and phosphorus (important in plant growth) as they decay;
- Absorbing moisture for growth and releasing it as vapor through transpiration;
- Preventing erosion by reducing the force of rainfall at the soil surface and by intercepting and absorbing water, rather than allowing it to run off directly;
- Harboring a diversity of wildlife;
- Acting as windbreaks;

Providing us with shade and beauty on a largely agricultural and urban landscape.

Forests and people are connected, and have been since ancient times. We have always had a special relationship based on survival. It was a delicate chain of existence that we once treated with respect and appreciation. But people began to upset this balance. They saw the forest not as a part of them but as something to be conquered. They used the seemingly limitless forest, cutting down millions of trees. But now it is coming to our attention that the forests *do* have limits and it is time to bring them back into balance.

All living things including marbled murrelets, tailed frogs and even fungi depend on forests. There are as many as 1,500 invertebrates living on and in one ancient forest tree! Some of these species may hold the key to unlocking scientific mysteries. Each plant and animal is unique and many of these animals depend entirely on the forests.

There is still much we don't know about forest ecosystems but each day is leading to new discoveries. Each animal, insect and plant contains its individual genetic material that has been evolving for thousands of years. Protecting the forests does not just mean saving a lot of trees, it is preserving a process of life that started billions of years ago. Ancient forests bring a better understanding of how forests work.

Forests protect our waters and manage our climate. When it rains in the forests, the leaves allow the water to slowly drip to the ground. When a forest has been clear-cut, the rain pours down hard on the unprotected soil. The dirt then washes into streams, muddying the waters. This is unhealthy for the fish, and can cause flooding. Also, without trees, the moisture in the air evaporates quickly, changing the climate of nearby forests. This process prevents trees from receiving the water they need.

Natural forests add to the economy. In 1988, six billion dollars were added to the economies of Washington and Oregon because of recreation in forests. People enjoy and appreciate fresh air, clear water, beautiful scenery and wildlife. So, places with all of the above are ideal tourist spots. Also, businesses like to be located in areas of such serene beauty.

Without the forests we would have much less oxygen. One acre of forest provides over 6 tons of oxygen per year. This is because trees (and all green plants) use a process called photosynthesis, during which they take in carbon dioxide and, as a by-product, produce oxygen. Plants "breathe" carbon dioxide, like we breathe oxygen. There has been a balance between species that breathe out carbon dioxide and take in oxygen, and species that take in carbon dioxide and exhale oxygen. Since the 1800's this balance has been upset. Fossil fuels, when burned, create carbon dioxide, so carbon dioxide levels have risen dramatically. Unfortunately, this gas, in large amounts, acts like an insulator and keeps heat near the surface of the Earth. This is called the "greenhouse effect."

Deforestration

Deforestation is the conversion of forested areas to non-forested land, for uses such as: pasture, urban use, logging purposes, and can result in arid land and wastelands. In many countries, deforestation is ongoing and is shaping climate and geography. Deforestation results from removal of trees without sufficient reforestation, and results in declines in habitat and biodiversity, wood for fuel and industrial use, and quality of life. Forests disappear naturally as a result of broad climate change, fire, hurricanes or other disturbances, however most deforestation in the past 40,000 years has been anthropogenic. Human induced deforestation may be accidental such as in the case of forests in Europe adversely affected by acid rain. Improperly applied logging, fuelwood collection, fire management or grazing can also lead to unintentional deforestation. However, most anthropogenic deforestation is deliberate. The consequences of deforestation are largely unknown and the impacts not verified by sufficient scientific data leading to considerable debate amongst scientists.

Deforestation is the permanent destruction of indigenous forests and woodlands. The term does not include the removal of industrial forests such as plantations of gums or pines. Deforestation has resulted in the reduction of indigenous forests to four-fifths of their pre-agricultural area. Indigenous forests now cover 21% of the earth's land surface.

Causes of Deforestration

There are two main causes of deforestation. The primary and most common reasons for deforestation are known as the direct causes. Logging, overpopulation, urbanization, dam construction etc are under direct causes. The other main cause of deforestation is known as natural causes since they are brought by the Mother Nature.

Rapid population growth has resulted to the conversion of forest areas to non-forest lands for settlement and farming. Together with this is urbanization and residential area expansion. This takes a significant loss of forest lands both for harvesting forest products as more people need more lumber to build their houses and for developing the greater area their houses, malls, business centers will be built.

An increase in population also means an increase in produce consumption. Thus, rainforests are destroyed and converted to cattle pasture to supply the burgeoning demand for meat. In Central America, almost half of the rainforests have been slashed and burned for cattle farming in order comply with foreign demands.

Twenty-five per cent of the Amazon's forests have also been destroyed for cattle ranches.

Lack of government legislation for land reforms has also cleared the forest especially in developing countries like of the South East Asian nations. People in that region are among the poorest in the world and are desperate for a piece of land. Unequal distribution of resources has led these people to find their way to exploit the forests.

Another reason that denudes the forest is exploitative economic development schemes and the powerlessness of government to safeguard its resources. Poor countries in their attempt to increase their revenues are in a way exploiting their resources like the forests. Timber is exported to reduce the national debt. Countries rich in mineral resources open their doors to multinational mining corporations that clear the forests as they go with their operations. The government especially those belonging in the Third World cannot curb commercial logging and implement a total log ban in exchange to higher foreign exchange rates. Development projects like dams, roads, and airports contracted by the government also cause deforestation.

While most causes of deforestation occur due to human activities, there are uncontrolled causes of deforestation such as forest fires, volcanic eruption, and typhoon.

Forest fires are started by lightning, and strong winds help to spread the flames. Drought in the forest has increased the amount of flammable bush and debris on the forest floor. Forest fires destroy immeasurable amount of valuable timber. They kill not only trees but also other living things.

Meanwhile, volcanic eruption is one of the several natural forces capable of causing damage to forests. The ashes emitted during the eruption coat tree leaves, which then interfere with photosynthesis. Animal population is also devastated. The organisms that survive have to cope with the changed habitat and reduced food supplies.

Last is typhoon. These are violent storms when fierce winds destroy much of the island's rain forest.

People can only hope that the uncontrollable forces causing deforestation would not do great damage. However, right decisions and good actions must be taken to address the problems brought by the other reasons of deforestation where the cause and the end result is at the hands of the people.

Effects of Deforestration

Deforestation is the process of converting forested lands into non-forest sites that are ideal for crop raising, urbanization and industrialization. Because deforestation is a serious concept, there are also serious effects to the surroundings.

Effects of deforestation can be classified and grouped into effects to biodiversity, environment and social settings. Because deforestation basically involves killing trees in forests, there are so many effects that can be enumerated as results of the activity.

When forests are killed, nature basically requires people to renew the forest. Reforestation is one concept that is in the opposite direction as deforestation, but is proven to be a much harder effort than deforestation.

So the rate of deforestation has not been offset by the rate of reforestation. Thus, the world is now in a troubled state when it comes to issues concerning the environment.

Pollution is rapidly growing along with population. Forests are greatly helping reduce the amount of pollutants in the air. So, the depletion of these groups of trees is greatly increasing the risk that carbon monoxide would reach the atmosphere and result in the depletion of the ozone layer, which in turn results to global warming.

Environment Change

One major effect of deforestation is climate change. Changes to the surroundings done by deforestation work in many ways. One, there is abrupt change in temperatures in the nearby areas. Forests naturally cool down because they help retain moisture in the air.

Second is the long process of global climate change. As mentioned above, deforestation has been found to contribute to global warming or that process when climates around the world become warmer as more harmful rays of the sun comes in through the atmosphere.

The ozone layer is a mass of oxygen or O_3 atoms that serves as shield in the atmosphere against the harmful ultraviolet rays from the sun. Because ozone is made up of oxygen atoms, oxygen react with carbon monoxide. Such reaction would use up oxygen atoms.

It follows that when there are more carbon monoxide atoms going to the atmosphere, the volume of oxygen would decline. Such is the case of ozone depletion.

The third effect to the environment would be on the water table underneath the ground. Water table is the common source of natural drinking water by people living around forests.

Water table is replenishing. That means, the supply of water underground could also dry up if not replenished regularly. When there is rain, forests hold much of the rainfall to the soil through their roots.

Thus, water sinks in deeper to the ground, and eventually replenishing the supply of water in the water table. Now, imagine what happens when there is not enough forests anymore. Water from rain would simply flow through the soil surface and not be retained by the soil.

Or other than that, the water from rain would not stay in the soil longer, for the process of evaporation would immediately set in. Thus, the water table is not replenished, leading to drying up of wells.

Effect to biodiversity

Forests are natural habitats to many types of animals and organisms. That is why, when there is deforestation, many animals are left without shelters. Those that manage to go through the flat lands and residential sites are then killed by people.

Through the years, it is estimated that there are millions of plant and extinct animal species that have been wiped out because they have been deprived of home. Thus, biodiversity is significantly lowered because of the savage deforestation practices of some people.

Wildlife advocates have been constantly reminding that several wild animals left in the world could still be saved if deforested forests would only be reforested and the practice of slash and burn of forests would be totally abandoned.

Social effects of deforestation

Deforestation is hardly hitting the living conditions of indigenous people who consider forests as their primary habitats. Imagine how they are rendered homeless when forests are depleted. These natives would be forced to live elsewhere, and are usually left to becoming mendicants in rural and urban areas.

Overall, effects of deforestation cannot be offset by the contribution of the practice to development. While it is logical that progress is very much needed by mankind, it must also be noted that nature knows no defeat. Destruct it and it would certainly retaliate, one way or another.

5

MINERAL RESOURCES

A **mineral** is a naturally occurring solid formed through geological processes that has a characteristic chemical composition, a highly ordered atomic structure, and specific physical properties. A rock, by comparison, is an aggregate of minerals and need not have a specific chemical composition. Minerals range in composition from pure elements and simple salts to very complex silicates with thousands of known forms. The study of minerals is called mineralogy.A mineral is an element or chemical compound that is normally crystalline and that has been formed as a result of geological processes.

Man's progress in civilization from the historic stage has been punctuated by his mastery over technology in extracting the mineral resources from different geological formations of diverse ages and in utilizing these mineral resources in various industries for the manufacture of essential products for human life. Economic edifice of any country depends upon is industrial growth, which in turn is dependable upon the availability of mineral resources of that country.

Differences between minerals and rocks

A mineral is a naturally occurring solid with a definite chemical composition and a specific crystalline structure. A rock is an aggregate of one or more minerals. (A rock may also include organic remains and mineraloids.) Some rocks are predominantly composed of just one mineral. For example, limestone is a sedimentary rock composed almost entirely of the mineral calcite. Other rocks contain many minerals, and the specific minerals in a rock can vary widely. Some minerals, like quartz, mica or feldspar are common, while others have been found in only four or five locations worldwide. The vast majority of the rocks of the Earth's crust consist of quartz, feldspar, mica, chlorite, kaolin, calcite, epidote, olivine, augite, hornblende, magnetite, hematite, limonite and a few other minerals.Over half of the mineral species known are so rare that they have only been found in a handful of samples, and many are known from only one or two small grains.

Commercially valuable minerals and rocks are referred to as industrial minerals. Rocks from which minerals are mined for economic purposes are referred to as ores.

Types Of Mineral

Minerals are broadly grouped into two categories.

(1) Metallic and

(2) Non-Metallic

Metallic minerals find their use in the manufacture of a small pin up to an aircraft and big vehicle. The *non-metallic* minerals are used in Tailoring trade, agriculture, and medicine and in big industries for the manufacture of various items of our daily use. Ever since man's emergence from the Neolithic, pursuit for metals and the development of metallurgy has been the keystone of his economic and industrial progress. Minerals thus form one of the modern world's important resource in support and fulfillment of human life. The wide range of their uses has made great exploitation on the world's stock of minerals and in some cases to such an extent that their accessibility stock in the earth's crust is reaching depletion point. The diminishing resource of minerals have called forth warning note from geologist and industrial economists. Researchers and technologists have made exhaustive efforts to minimize the consumption of metals and to use non- mineral products such as plastic in place of metals so that the exponential rate at which our civilization is consuming the land based mineral resources be put off.

Mineral Fuels

Mineral fuels are fossil source fuels, that is, carbon or hydrocarbons found in the earth's crust. Fossil fuels range from volatile materials with low carbon:hydrogen ratios like methane, to liquid petroleum to nonvolatile materials composed of almost pure carbon, like anthracite coal. Methane can be found in hydrocarbon fields, alone, associated with oil, or in the form of methane clathrates. It is generally accepted that they formed from the fossilized remains of dead plants and animals by exposure to heat and pressure in the Earth's crust over hundreds of millions of years.

Fossil fuels are non-renewable resources because they take millions of years to form, and reserves are being depleted much faster than new ones are being formed. Concern about fossil fuel supplies is one of the causes of regional and global conflicts. The production and use of fossil fuels raise environmental concerns. A global movement towards the generation of renewable energy is therefore under way to help meet increased energy needs.

The burning of fossil fuels produces around 21.3 billion tons (21.3 gigatons) of carbon dioxide per year, but it is estimated that natural processes can only absorb about half of that amount, so there is a net increase of 10.65 billion tones of atmospheric carbon dioxide per year (one tonne of atmospheric carbon is equivalent to 44/12 or 3.7 tons of carbon dioxide).Carbon dioxide is one of the greenhouse gases that enhances radiative forcing and contributes to global warming, causing the average surface temperature of the Earth to rise in response, which climate scientists agree will cause major adverse effects, including reduced biodiversity.

Mineral Resources in India

The main mineral resources present in india are Coal, Limestone, Bauxite and Iron ore.The description of these mineral resources are following below:

COAL : The basis source of energy for the industrial development is the coal, the solid or fossil fuel as it is called and the district Rajouri has Kalakote, Methka, Moghla, Chokkar the main coal. The coal is of semi-anthracite having fixed carbon for percentage from 79 to 80, which is related to Himalayan. The intense and complex tectonic movement has restricted the economical limits of these coalfields. However, these can meet the increasing demand of local consumption and industrial development of Jammu region.

LIMESTONE : Limestone is an important raw material used in the manufacture of cement and in chemical, metallurgical, paper making glass manufacture and sugar refining. It is also used in bleaching powder and caustic soda. Extensive deposits of limestone associated with carboniferous sequence of Kalakote, Methka and Thannamandi area of Rajouri district can be placed in any of its above industrial use only after conducting systematic geological investigation.

BAUXITE : It is good source of aluminium and can be used for extraction of aluminium metal. It can be used for refractory purposes and also for the extraction of gallium. It is a blanket type of deposit occurring unconformable over the great limestone of Kalakote, Methka, Moghla and Chakkar area. It is generally dark gray to cream in color and exhibited diasporic variety.

IRON ORE : Though district Rajouri is devoid of any significant occurrence of iron ores but in Gagrot-Khandli devi area iron ore occurs in a lenticular band associated with carbonaceous and fessuginous shake and slate.

BENTONITE : It is clay type of deposits usually shows whitish color and gives soapy touch. It is used in oil well drilling, oil refining, paints, sealing in reservoirs and irrigational canals. It also finds its use in agricultural spray insecticides. Thick bands of bentonite are located in three different places near Budhal. In addition to these minerals resources present in Rajouri district, the Panjal trap has a massive volcanic rock, which forms the northern and northwestern. Envisions of Rajouri city can be used in building purpose. With the increasing demand of the stone industry, polished slabs of Panjal traps can be manufactured after arriving at their physical mineralogical and textural properties.

MURREE : Murree formation supposed to be the main source of hydrocarbon is predominantly developed in this area. Unfortunately success in exploring the oil and natural gas commission workers have not achieved petroleum in saruinsar area due to certain technical problems, extensive efforts are being made by these workers in Rajouri district also to locate sports for exploration of petroleum in future.

Effects of Extracting ore of Mineral Resources

Mining is the extraction of valuable minerals or other geological materials from the earth, usually from an ore body, vein or (coal) seam. Materials recovered by mining include base metals, precious metals, iron, uranium, coal, diamonds, limestone, shale oil; rock salt and potash. Any material that cannot be grown through agricultural processes, or created artificially in a laboratory or factory, is usually mined. Mining in a wider sense comprises extraction of any non-renewable resource (e.g., petroleum, natural gas, or even water).

Modern mining processes involve prospecting for ore bodies, analysis of the profit potential of a proposed mine, extraction of the desired materials and finally reclaimation of the land to prepare it for other uses once the mine is closed. The nature of mining processes creates a potential negative impact on the environment both during the mining operations and for years after the mine is closed. This impact has led to most of the world's nations adopting regulations to moderate the negative effects of mining operations. Safety has long been a concern as well, though modern practices have improved safety in mines significantly. Mining today is able to profitably and safely recover minerals with little negative impact to the environment.

Environmental issues can include erosion, formation of sink holes, loss of biodiversity, and contamination of groundwater and surface water by chemicals from the mining process. Modern mining companies in some countries are required to follow environmental and rehabilitation codes, ensuring the area mined is returned to close to its original state. In some countries with pristine environments, such as large parts of Australia, this is impossible despite the best intentions. Some mining methods have devastating environmental and public health effects.

Mining can have adverse effects on surrounding surface and ground water if protective measures are not taken. The result can be unnaturally high concentrations of some chemicals, such as arsenic and sulfuric acid, over a significant area of surface or subsurface. There is potential for massive contamination of the area surrounding mines due to the various chemicals used in the mining process as well as the potentially damaging compounds and metals removed from the ground with the ore. Large amounts of water produced from mine drainage, mine cooling, aqueous extraction and other mining processes increases the potential for these chemicals to contaminate ground and surface water. In well-regulated mines, hydrologists and geologists take careful measurements to mitigate any type of water contamination that could be caused by the mine's operations. In modern American mining, operations must, under federal and state law, meet standards for protecting surface and ground water from contamination, including **acid mine drainage** (AMD). To mitigate these problems water is continuously monitored. The five principal technologies used to control water flow at mine sites are diversion systems, containment ponds, groundwater pumping systems, subsurface drainage systems, and subsurface barriers. In the case of AMD, contaminated water is generally pumped to a treatment facility that neutralizes the contaminants.Dissolution and transport of metals and heavy metals by run-off and ground water is another example of environmental problems with mining.

6

ENERGY RESOURCES

Wind is simple air in motion. It is caused by the uneven heating of the earth's surface by the sun. Since the earth's surface is made of very different types of land and water, it absorbs the sun's heat at different rates. Wind is the flow of air or other gases that compose an atmosphere (including, but not limited to, the Earth's). In short terms-wind is air molecules in motion. Winds are commonly classified by their spatial scale, their speed, the types of forces that cause them, the geographic regions in which they occur, and their effect. While wind is often a standalone weather phenomenon, it can also occur as part of a storm system, most notably in a cyclone.

During the day, the air above the land heats up more quickly than the air over water. The warm air over the land expands and rises, and the heavier, cooler air rushes in to take its place, creating winds. At night, the winds are reversed because the air cools more rapidly over land than over water. In the same way, the large atmospheric winds that circle the earth are created because the land near the earth's equator is heated more by the sun than the land near the North and South Poles.

Today, wind energy is mainly used to generate electricity. Wind is called a renewable energy source because the wind will blow as long as the sun shines.

Bioenergy

Bioenergy is renewable energy made available from materials derived from biological sources. In its most narrow sense it is a synonym to biofuel, which is fuel derived from biological sources. In its broader sense it includes biomass, the biological material used as a biofuel, as well as the social, economic, scientific and technical fields associated with using biological sources for energy. This is a common misconception, as bioenergy is the energy extracted from the biomass, as the biomass is the fuel and the bioenergy is the energy contained in the fuel.

Biomass is any organic material which has stored sunlight in the form of chemical energy. As a fuel it may include wood, wood waste, straw, manure, sugar cane, and many other byproducts from a variety of agricultural processes.

The conventional resources of energy such as fuel oil, gas and coal are finite and the already depleting very rapidly. Being our country an agricultural one and more than 70 % of the Indian

Population living in the villages, about 80% of their energy need is for domestic purpose (i.e. for cooking, lighting, water pumping etc.) is fulfilled by non commercial sources of energy like cow dung, firewood, agricultural waste etc.

The BIOMASS / ORGANIC WASTE generated in our country (villages in urban & urban areas) consists of following types :

Bio-mass : Crop residues / WasteAgricultural waste Animal Dung

OrganicWaste : Agro – food processing industries Pulp & paper Edible & non edible oil, exploring units waste, Sugar/ Distilleries, Industrial wastes, Municipal waste, Industrial waste

The above wide range of industrial, urban & agricultural & crop residues and animal dung etc waste which organic waste by nature is proved to be the best and effective source of energy generation. The energy produced from such **Bio-mass or organic waste** is well known as **BIO-ENERGY.**

The conventional resources of energy such as fuel oil, gas and coal are finite and the already depleting very rapidly, Being our country an agricultural one and more than 70 % of the Indian Population living in the villages, about 80% of their energy need is for domestic purpose (i.e. for cooking, lighting, water pumping etc. is fulfilled by non commercial sources of energy like cow dung, firewood, agricultural waste etc.

Geothermal Energy

Geothermal power (from the Greek roots *geo*, meaning earth, and *thermos*, meaning heat) is energy generated from heat stored in the earth, or the collection of absorbed heat derived from underground. Prince Piero Ginori Conti tested the first geothermal generator on 4 July 1904, at the Larderello dry steam field in Italy. The largest group of geothermal power plants in the world is located at The Geysers, a geothermal field in California, United States.The Philippines and Iceland are the only countries to generate a significant percentage of their electricty from geothermal sources;

in both countries 15-20% of power comes from geothermal plants. As of 2008, geothermal power supplies less than 1% of the world's energy. The most common type of geothermal power plants (binary plants) are closed cycle operations and release essentially no GHG emissions; geothermal power is available 24 hours a day with average availabilities above 90% (compared to about 75% for coal plants).

The word **geothermal** comes from the Greek words geo (earth) and therme (heat). So, geothermal energy is heat from within the earth. We can use the steam and hot water produced inside the earth to heat buildings or generate electricity. Geothermal energy is a **renewable** energy source because the water is replenished by rainfall and the heat is continuously produced inside the earth.

Energy Inside the Earth

Geothermal energy is generated in the earth's core, about 4,000 miles below the surface. Temperatures hotter than the sun's surface are continuously produced inside the earth by the slow decay of radioactive particles, a process that happens in all rocks. The earth has a number of different layers:

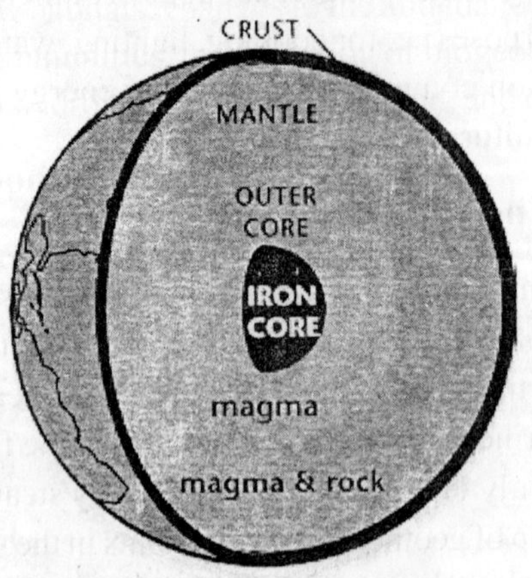

THE EARTH'S INTERIOR

The core itself has two layers: a **solid iron core** and an outer core made of very hot melted rock, called **magma**.

The **mantle** which surrounds the core and is about 1,800 miles thick. It is made up of magma and rock.

The **crust** is the outermost layer of the earth, the land that forms the continents and ocean floors. It can be three to five miles thick under the oceans and 15 to 35 miles thick on the continents.

The earth's crust is broken into pieces called **plates**. Magma comes close to the earth's surface near the edges of these plates. This is where volcanoes occur. The lava that erupts from volcanoes is partly magma. Deep underground, the rocks and water absorb the heat from this magma. The temperature of the rocks and water get hotter and hotter as you go deeper underground.

People around the world use geothermal energy to heat their homes and to produce electricity by digging deep wells and pumping the heated underground water or steam to the surface. Or, we can make use of the stable temperatures near the surface of the earth to heat and cool buildings.

Advantages of Geothermal Power

Geothermal power requires no fuel, and is therefore virtually emissions free and insusceptible to fluctuations in fuel cost. And because a geothermal power station doesn't rely on transient sources of energy, unlike, for example, wind turbines or solar panels, its capacity factor can be quite large; up to 90% in practice.

Geothermal has minimal land use requirements; existing geothermal plants use 1-8 acres per megawatt (MW) versus 5-10 acres per MW for nuclear operations and 19 acres per MW for coal power plants. It also offers a degree of scalability: a large geothermal plant can power entire cities while smaller power plants can supply more remote sites such as rural villages

Geothermal energy is the heat from the Earth. It's clean and sustainable. Resources of geothermal energy range from the shallow ground to hot water and hot rock found a few miles beneath the Earth's surface, and down even deeper to the extremely high temperatures of molten rock called magma.

Almost everywhere, the shallow ground or upper 10 feet of the Earth's surface maintains a nearly constant temperature between 50° and 60°F (10° and 16°C). Geothermal heat pumps can tap into this resource to heat and cool buildings. A geothermal heat pump system consists of a heat pump, an air delivery system (ductwork), and a heat exchanger-a system of pipes buried in the shallow ground near the building. In the winter, the heat pump removes heat from the heat exchanger and pumps it into the indoor air delivery system. In the summer, the process is reversed, and the heat pump moves heat from the indoor air into the heat exchanger. The heat removed from the indoor air during the summer can also be used to provide a free source of hot water.

Tidal and Ocean Energy

Tidal power is the only form of energy which derives directly from the relative motions of the Earth–Moon system, and to a lesser extent from the Earth–Sun system. The tidal forces produced by the Moon and Sun, in combination with Earth's rotation, are responsible for the generation of the tides. Other sources of energy originate directly or indirectly from the Sun, including fossil fuels, conventional hydroelectric, wind, biofuels, wave power and solar. Nuclear is derived using radioactive material from the Earth, geothermal power uses the heat of magma below the Earth's crust, which comes from radioactive decay.

Tidal energy is generated by the relative motion of the Earth, Sun and the Moon, which interact via gravitational forces. Periodic changes of water levels, and associated tidal currents, are due to the gravitational attraction by the Sun and Moon. The magnitude of the tide at a location is the result of the changing positions of the Moon and Sun relative to the Earth, the effects of Earth rotation, and the local shape of the sea floor and coastlines.

Because the Earth's tides are caused by the tidal forces due to gravitational interaction with the Moon and Sun, and the Earth's rotation, tidal power is practically inexhaustible and classified as a renewable energy source.

A tidal energy generator uses this phenomenon to generate energy. The stronger the tide, either in water level height or tidal current velocities, the greater the potential for tidal energy generation.

Tidal movement causes a continual loss of mechanical energy in the Earth–Moon system due to pumping of water through the natural restrictions around coastlines, and due to viscous dissipation at the seabed and in turbulence. This loss of energy has caused the rotation of the Earth to slow in the 4.5 billion years since formation. During the last 620 million years the period of rotation has increased from 21.9 hours to the 24 hours we see now; in this period the Earth has lost 17% of its rotational energy. While tidal power may take additional energy from the system, increasing the rate of slowdown, the effect would be noticable over millions of years only, thus being negligable.

Tidal and wave technologies convert the kinetic energy of moving water into electricity. Ocean thermal systems tap the solar heat absorbed by marine waters to generate clean energy. In theory, these three ocean-based renewable resources could meet the world's energy requirements many times over, but they remain largely undeveloped at present.

Most tidal current technologies employ a turbine to transform kinetic energy into electricity.

Tidal Energy

Tidal energy takes advantage of the daily flow of tides and of more localized examples of water in motion. The gravitational pull of the moon drives tidal flows, while persistent currents and large-scale circulations such as the Gulf Stream are influenced by solar heating, water chemistry, and other factors. Most tidal current technologies employ a turbine to transform kinetic energy into electricity.

Two main types of tidal energy systems exist. Emerging tidal current technology does not alter the natural flow of water, whereas tidal barrages operate like conventional hydropower facilities.

Wave Energy

Waves are powered largely by the wind and the tides. The majority of the more than 1,000 patents relating to wave energy technology are for devices that harness kinetic energy associated with the up and down motion of the water column as waves pass through it.

Hundreds of patents exist for wave energy technology, and a variety of prototype devices have been tested. However, only one commercial wave power station has been developed a 500 kW plant, sited on the shoreline in Islay , Scotland . Demonstration projects are being examined for various locations in New England.

Ocean Thermal Energy

Ocean thermal energy conversion (OTEC) systems exploit temperature differences between warmer, surface layers and colder, deep layers of the ocean. All OTEC designs require a large-diameter intake pipe to pump cold water to the surface. There, they employ a variety of heat-exchange cycles to drive a turbine and generate electricity. OTEC technology is further from commercial application than wave and tidal systems. Small-scale OTEC units and individual system components have been tested successfully in waters off the coast of Hawaii.

Tapping ocean energy resources provides Massachusetts with opportunities to reduce reliance on fossil fuels and other conventional sources without producing pollution or greenhouse gases.

Ocean energy covers a series of emerging technologies that use the power of ocean currents, waves, and tides to create energy. While very few of these technologies have been implemented on a commercial scale, they show much promise for future development. asoline), but the lowest energy content by volume (about four times less than gasoline). It is the lightest element, and it is a gas at normal temperature and pressure.

Hydrogen Energy

Hydrogen is the simplest element known to man. Each atom

of hydrogen has only one proton. It is also the most plentiful gas in the universe. Stars are made primarily of hydrogen.

The sun is basically a giant ball of hydrogen and helium gases. In the sun's core, hydrogen atoms combine to form helium atoms. This process—called fusion—gives off radiant energy.

This **radiant energy** sustains life on earth. It gives us light and makes plants grow. It makes the wind blow and rain fall. It is stored as chemical energy in fossil fuels. Most of the energy we use today came from the sun's radiant energy.

Hydrogen gas is lighter than air and, as a result, it rises in the atmosphere. This is why hydrogen as a gas (H_2) is not found by itself on earth. It is found only in compound form with other elements. Hydrogen combined with oxygen, is water (H_2O). Hydrogen combined with carbon, forms different compounds such as methane (CH_4), coal, and petroleum. Hydrogen is also found in all growing things—biomass. It is also an abundant element in the earth's crust.

Hydrogen has the highest energy content of any common fuel by weight(about three times more than gasoline), but the lowest energy content by volume (about four times less than gasoline). It is the lightest element, and it is a gas at normal temperature and pressure.

Hydrogen has the highest energy to weight ratio, NASA has used it as a rocket fuel since the 1940's. Most people don't notice it, but if you watch a shuttle lift off you can see that the three rockets attached to the shuttle its self have a light blue almost clear flame, that is the on board hydrogen rocket. NASA also uses hydrogen for it's primary fuel while out in space, and for making drinking water, one pound of hydrogen when combined with oxygen will make nine pounds of pure distilled drinking water. Through the process, it will generate a significant amount of usable electricity as a byproduct. The Navy has been using electrolyzers for their submarines to make oxygen for long missions; they would turn on the Diesel engines, and turn the sea water into hydrogen and oxygen, essentially opposite of what NASA is doing. In fact the American Hydrogen Association has one of the original electrolyzers from a submarine. It is still fully functional today, and has over one hundred million life support hours to its credit since 1955, as a oxygen generator for submarines.

Hydrogen has many practical uses, for example, you can easily convert any combustion engine to run on hydrogen. Hydrogen can be used as a cooking fuel, to heat your home, drive your car, and mow your lawn. Hydrogen can run your generator and run the electricity for your home. With the addition of a fuel cell, hydrogen can be turned back to electricity to run your computer, your lights. It can be used in place of electricity, in place of gasoline, and in place or propane or natural gas (Camp). It can be used to suit all the world's power needs. Unlike with so many things, which only the rich more developed countries, can afford. Hydrogen can power any country where the sun shines.

Fossil Fuels

Fossil fuels or **mineral fuels** are fossil source fuels, that is, carbon or hydrocarbons found in the earth's crust. Fossil fuels range from volatile materials with low carbon:hydrogen ratios like methane, to liquid petroleum to nonvolatile materials composed of almost pure carbon, like anthracite coal. Methane can be found in hydrocarbon fields, alone associated with oil, or in the form of methane clathrates. It is generally accepted that they formed from the fossilized remains of dead plants and animals by exposure to heat and pressure in the Earth's crust over hundreds of millions of years

Fossil fuels are non-renewable resources because they take millions of years to form, and reserves are being depleted much faster than new ones are being formed. Concern about fossil fuel supplies is one of the causes of regional and global conflicts. The production and use of fossil fuels raise environmental concerns. A global movement toward the generation of renewable energy is therefore under way to help meet increased energy needs.

The burning of fossil fuels produces around 21.3 billion tons (21.3 gigatons) of carbon dioxide per year, but it is estimated that natural processes can only absorb about half of that amount, so there is a net increase of 10.65 billion tones of atmospheric carbon dioxide per year (one tonne of atmospheric carbon is equivalent to 44/12 or 3.7 tons of carbon dioxide). Carbon dioxide is one of the greenhouse gases that enhances radiative forcing and contributes to global warming, causing the average surface temperature of the Earth to rise in response, which climate scientists agree will cause major adverse effects, including reduced biodiversity.

Fossil fuels are formed from the preserved remains of organisms including phytoplankton and zooplankton that have settled to the sea (or lake) bottom in large quantities under anoxic conditions. Over geological time, this organic matter mixed with mud, is buried under heavy layers of sediment. The resulting high levels of heat and pressure cause the organic matter to chemically alter, first into a waxy material known as kerogen which is found in oil shales, and then with more heat into liquid and gaseous hydrocarbons in a process known as catagenesis.

Fossil fuels are of great importance because they can be burned (oxidized to carbon dioxide and water), producing significant amounts of energy.

The use of coal as a fuel predates recorded history. Coal was used to run furnaces for the melting of metal ore. Semi-solid hydrocarbons from seeps were also burned in ancient tmes but these materials were mostly used for waterproofing.

Commercial exploitation of petroleum, largely as a replacement for oils from animal sources (notably whale oil) for use in oil lamps began in the nineteenth century.

Natural gas, once flared-off as an un-needed byproduct of petroleum production, is now considered a very valuable resource.

Heavy crude oil, which is very much more viscous than conventional crude oil, and tar sands, where bitumen is found mixed with sand and clay, are becoming more important as sources of fossil fuel. Oil shale and similar materials are sedimentary rocks containing kerogen, a complex mixture of high-molecular weight organic compounds, which yield synthetic crude oil when heated (pyrolyzed). These materials have yet to be exploited commercially.

These fuels are employed in internal combustion engines, fossil fuel power stations and other uses.

Prior to the latter half of the eighteenth century, windmills or watermills provided the energy needed for industry such as milling flour, sawing wood or pumping water, and burning wood or peat provided domestic heat. The wide-scale use of fossil fuels, coal at first and petroleum later, to fire steam engines, enabled the Industrial Revolution. At the same time, gas lights using natural gas or coal gas were coming into wide use. The invention of the internal combustion engine and its use in automobiles and trucks greatly increased the demand for gasoline and diesel oil, both made from fossil fuels. Other forms of transportation, railways and aircraft also required fossil fuels. The other major use for fossil fuels is in generating electricity.

Fossil fuels are also the main source of raw materials for the petrochemical industry.

Fossil fuels, coal, oil and natural gas, are a non-renewable source of energy. Formed from plants and animals that lived up to 300 million years ago, fossil fuels are found in deposits beneath the earth. The fuels are burned to release the chemical energy that is stored within this resource.

Types of Fossil Fuels

Fossil fuels are responsible for providing the energy needed worldwide for many household and industrial purposes today. Deposits of fossil fuels are found throughout the world deep inside the earth. Fossil fuels are the carbon rich remains of ancient vegetation and other organisms that have endured intense heat and pressure inside the earth over periods of millions of years. Severe heat and pressure over time convert these organic remains into fossil fuels, which collect in reservoirs that are sought after by oil focused companies such as Western Pipeline Corporation as well as investors for extraction. Though quite a few types of fossil fuels exist, here we examine three main types that are widely utilized for energy production and many other products: petroleum, coal and natural gas.

Though quite a few types of fossil fuels exist, here we examine three main types that are widely utilized for energy production and many other products: petroleum, coal and natural gas.

Petroleum: Also called crude oil, the term petroleum encompasses multiple types of hydrocarbons, which are compounds consisting primarily of hydrogen and carbon but possibly containing other elements as well. Petroleum forms mainly from marine vegetation and bacteria that lived in the oceans or other saltwater environments millions of years ago. Petroleum deposits are often found in the same locations as natural gas, each of which can be extracted for energy production. Petroleum is used in the production of plastic and medications among many other products.

Coal : Coal forms from plants such as ferns, moss and trees which lived near shorelines and in swamps and bogs millions of years ago. When these plants die, they are slowly covered with sediment and over time pressed deep into the earth where they are affected by mounting heat and pressure. Under these conditions, the organic matter becomes richer in carbon and hydrogen, and increasingly deprived of oxygen. Coal goes through various stages of development based on its increasing carbon content, and coal containing higher levels of carbon burns cleaner than those with lower levels. The purest form of coal is graphite, which consists almost entirely of carbon.

Natural Gas : Natural gas forms mainly from the remains of plankton, or a type of small water organisms including algae. Consisting mostly of methane, natural gas is often found on top of deposits of petroleum due to its lower density, and is extracted in the same process. However, deposits containing only natural gas do exist. Natural gas is desirable in part because it burns cleaner than coal and petroleum. Natural gas is commonly used in residential applications for home heating and has a myriad of other applications.

Fossil fuels are produced from animals and plants that were living 300 million years back. Oil, gas, and coal are the common examples of fossil fuels. The term "fossil fuel" is used to describe these fuels because their source dates back to the prehistoric period. These fuels are mostly obtained in the form of deposits under the earth. Fossil fuels have chemical energy stored within them. When these fuels are burnt, the stored energy gets released and serves the purpose. Fossil fuels are used to meet around 85% of the total energy requirement of the world.

Origin of Fossil Fuels: An Overview

The biogenic theory gives an idea about the origin of fossil fuels. Algae and zooplankton that settled under the water bodies like sea or huge lake in the prehistoric period are the main source of petroleum. Anoxic conditions are required for the formation of petroleum. On the other hand, terrestrial plants that died in prehistoric period are the main source of coal.

Use of Fossil Fuels

Fossil fuels assume huge importance in the energy scenario of the world. Oil alone meets around 40% of the total requirement of energy. The distillate fuels derived from crude oil are extensively used for transportation, cooking, industrial production and many other purposes. Heating oil, turbo-jet fuel, kerosene, and diesel are some of the examples of distillate fuels. Fossil fuels constitute the chief sources of electrical energy. The thermal power plants make use of coal for power generation. The power plants that use coal contribute 50% of the total demand of energy across the world.

Importance of Sustainable Energy

Reducing dependency on fossil fuels is a major challenge for most economically advanced countries of the world as there is a very important link between the usage of fossil fuels and the economic conditions of individual. One of the major reasons behind the increasing demand to reduce the reliance on fossil fuels is the fact that they are expected to run out at a certain point of time in the future. This is where sustainable energy is expected to be useful as it would last longer than any other form of fuel source.

Ways of Reducing Dependency on Fossil Fuels

There are certain ways in which the Dependency on the fossil fuels may be reduced. One major way is to change or replace the technologies that are used in the construction of the automobiles as they are the major sources of air pollution. Experts are of the opinion that people need to use electric batteries in their cars. The principal reason behind this argument is that this using battery driven cars would reduce the Dependency on the fossil fuels.

Experts Opinion on Reducing Dependency on Fossil Fuels

Experts from major oil companies have been instructing people to use alternative fuels. They have suggested a number of steps that may be followed in order to reduce the dependency on fossil fuels. Those steps may be suggested as below:

- Using diesel in the cars
- Turning lawns into organic gardens
- Using solar energy at home
- Using wood for the purposes of heating
- Using telecommunications in the professional life

People connected to the fields of science and technologies have been coming up with new gases. In the United States of America the scientists have been busy in creating new gases like syngas, which is a combination of hydrogen and carbon oxides. These are supposed to replace gases like ethanol, which is supposed to have lesser environmental value. Biomass would be important in this context.

Advantages of Reducing Dependency on Fossil Fuels

Reducing dependency on fossil fuels has its economic advantages as well. It has been calculated that the engines operated with electricity are far more efficient in terms of mileage than the vehicles that run on fossil fuels like petroleum for example. It has been assumed that any possible increase in the generation of electricity is supposed to be a combination of several types of alternative energy resources. Those may be mentioned as below:

- Hydroelectricity
- Nuclear energy
- Concentrated solar thermal energy
- Wind power
- Geothermal energy
- Solar cells

Nuclear Power

Nuclear power is any nuclear technology designed to extract usable energy from atomic nuclei via controlled nuclear reactions. The most common method today is through nuclear fission, though other methods include nuclear fusion and radioactive decay. All utility-scale reactors heat water to produce steam, which is then converted into mechanical work for the purpose of generating electricity or propulsion. In 2007, 14% of the world's electricity came from nuclear power. More than 150 nuclear-powered naval vessels have been built, and a few radioisotope rockets have been produced.

Nuclear power is generated using Uranium, which is a metal mined in various parts of the world.

The first large-scale nuclear power station opened at Calder Hall in Cumbria, England, in 1956.

Some military ships and submarines have nuclear power plants for engines.

Nuclear power produces around 11% of the world's energy needs, and produces huge amounts of energy from small amounts of fuel, without the pollution that you'd get from burning fossil fuels.

The main bit to remember

Nuclear power stations work in pretty much the same way as fossil fuel-burning stations, except that a "chain reaction" inside a nuclear reactor makes the heat instead. The reactor uses Uranium rods as fuel, and the heat is generated by **nuclear fission:** neutrons smash into the nucleus of the uranium atoms, which split roughly in half and release energy in the form of heat. Carbon dioxide gas or water is pumped through the reactor to take the heat away, this then heats water to make steam. The steam drives turbines which drive generators.

Nuclear reactor

Modern nuclear power stations use the same type of turbines and generators as conventional power stations.

In Britain, nuclear power stations are often built on the coast, and use sea water for cooling the steam ready to be pumped round again. This means that they don't have the huge "cooling towers" seen at other power stations.

The reactor is controlled with "control rods", made of boron, which absorb neutrons. When the rods are lowered into the reactor, they absorb more neutrons and the fission process slows down. To generate more power, the rods are raised and more neutrons can crash into uranium atoms.

Nuclear Energy is released by the splitting (fission) or merging together (fusion) of the nuclei of atom(s). The conversion of nuclear mass to energy is consistent with the mass-energy equivalence formula $''E = ''m.c^2$, in which $''E$ = energy release, $''m$ = mass defect, and c = the speed of light in a vacuum (a physical constant). Nuclear energy was first discovered by French physicist Henri Becquerel in 1896, when he found that photographic plates stored in the dark near uranium were blackened like X-ray plates, which had been just recently discovered at the time 1895.

Nuclear chemistry can be used as a form of alchemy to turn lead into gold or change any atom to any other atom (albeit through many steps). Radionuclide (radioisotope) production often involves irradiation of another isotope (or more precisely a nuclide), with alpha particles, beta particles, or gamma rays. Iron has the highest binding energy per nucleon of any atom. If an atom of lower average binding energy is changed into an atom of higher average binding energy, energy is given off. The chart shows that fusion of hydrogen, the combination to form heavier atoms, releases energy, as does fission of uranium, the breaking up of a larger nucleus into smaller parts. Stability varies between isotopes: the isotope U-235 is much less stable than the more common U-238.

Nuclear energy is released by three *exoenergetic* (or exothermic) processes:

- Radioactive decay, where a neutron or proton in the
- radioactive nucleus decays spontaneously by emitting either particles,
- electromagnetic radiation (gamma rays),
- neutrinos (or all of them)
- Fusion, two atomic nuclei fuse together to form a heavier nucleus
- Fission, the breaking of a heavy nucleus into two (or more rarely three) lighter nuclei

Changes can occur in the structure of the nuclei of atoms. These changes are called nuclear reactions. Energy created in a nuclear reaction is called nuclear energy, or atomic energy.

Nuclear energy is produced naturally and in man-made operations under human control.

- Naturally: Some nuclear energy is produced naturally. For example, the Sun and other stars make heat and light by nuclear reactions.
- Man-Made: Nuclear energy can be man-made too. Machines called nuclear reactors, parts of nuclear power plants, provide electricity for many cities. Man-made nuclear reactions also occur in the explosion of atomic and hydrogen bombs.
- Nuclear energy is produced in two different ways, in one, large nuclei are split to release energy. In the other method, small nuclei are combined to release energy.

  For a more detailed look at nuclear fission and nuclear fusion.
- Nuclear Fission: In nuclear fission, the nuclei of atoms are split, causing energy to be released. The atomic bomb and nuclear reactors work by fission. The element uranium is the main fuel used to undergo nuclear fission to produce energy since it has many favorable properties. Uranium nuclei can be easily split by shooting neutrons at them. Also, once a uranium nucleus is split, multiple neutrons are released which

are used to split other uranium nuclei. This phenomenon is known as a chain reaction.

Nuclear Fusion: In nuclear fusion, the nuclei of atoms are joined together, or fused. This happens only under very hot conditions. The Sun, like all other stars, creates heat and light through nuclear fusion. In the Sun, hydrogen nuclei fuse to make helium. The hydrogen bomb, humanity's most powerful and destructive weapon, also works by fusion. The heat required to start the fusion reaction is so great that an atomic bomb is used to provide it. Hydrogen nuclei fuse to form helium and in the process release huge amounts of energy thus producing a huge explosion.

Thermal Energy

The ultimate source of thermal energy available to mankind is the sun, the huge thermo-nuclear furnace that supplies the earth with the heat and light that are essential to life. The nuclear fusion in the sun increases the sun's thermal energy. Once the thermal energy leaves the sun (in the form of radiation) it is called heat. Heat is thermal energy *in transfer*. Thermal energy is part of the overall *internal energy* of a system.

Definitions

To note, $U_{thermal}$ is almost never the total energy of a system; for instance, there can be static energy that doesn't change with temperature, such as bond energy or rest energy ($E=mc^2$).

Other definitions

Thermal Energy comes from the movement of atoms and molecules in matter.

Thermal energy per particle is also called the average translational kinetic energy possessed by free particles given by equipartition of energy.

Thermal energy is the difference between the internal energy of an object and the amount that it would have at absolute zero. It includes the quantity of kinetic energy due to the motion of the internal particles of an object, and is increased by heating and reduced by cooling.

In a monatomic ideal gas, the thermal energy is exactly given by the kinetic energy of the constituent particles.

Energy that comes from heat. This kind of energy can be either kinetic or potential.Temperature is the measure of the average kinetic energy in the particles in an object. You use temperature instead of hot and cold. Temperature depends on the kinetic energy of particles. Particles have kinetic energy because they are in motion the faster moving the more energy. The more kinetic energy the higher temperature. An objects temperature is a measure of average kinetic energy. the temperature of a substance is not by how much you have. What is heat- Heat is a transfer of energy. Heat can be hot or cold. Thermal energy is the total energy of the particles that make up a substance. An object with a high temperature has a higher thermal energy than a one with a low temperature. States of matter- Are physical forms in which a substance can exist. Example- solid liquid gas.

At a more basic level, thermal energy comes form the movement of atoms and molecules in matter. It is a form of kinetic energy produced from the random movements of those molecules. Thermal energy of a system can be increased or decreased.

At a more basic level, thermal energy comes form the movement of atoms and molecules in matter. It is a form of kinetic energy produced from the random movements of those molecules. Thermal energy of a system can be increased or decreased.

When you put your hand over a hot stove you can feel the heat. You are feeling thermal energy *in transfer*. The atoms and molecules in the metal of the burner are moving very rapidly because the electrical energy from the wall outlet has increased the thermal energy in the burner. We all know what happens when we rub our hands together. Our mechanical energy increases the thermal energy content of the atoms in our hands and skin. We then *feel* the consequence of this - heat.

Thermal Energy Example

Take a small piece of ice out of your fridge and hold it in your hand. The thermal energy content of your hand is *higher* then the thermal energy content of the ice cube. The atoms that comprise your hand are moving more rapidly then the atoms that make up the ice cube. Therefore, there will be a transfer of thermal energy from your hand to the ice cube. While this thermal energy is *in transfer*, it is called heat. This will cause the atoms in the ice cube to *speed up* while the atoms in your hand *slow down*. The increase in speed of the ice cube atoms changes the state of water from solid to liquid. This transfer of thermal energy will continue until an equilibrium is reached between your hand, the ice (now water), and the air in the room.

Laws of Thermodynamics : When talking about thermal energy we must also talk about the laws of thermodynamics which express the laws of the interaction of energy and matter.

First law of thermodynamics: Energy and matter are interchangable but cannot be created or destroyed. The total amount of energy in the whole universe remains constant, only changing from one form to another.

Second Law of thermodynamics: This law states that any system always tends to move toward its probable state of energy. For example, a spring watch will run until the potential energy in the spring is used up. If no new energy is input into it (in the form of winding the spring up) then it has returned to its most probable state, which is really not to run. The most misunderstood law.

Third Law of Thermodynamics: This law is a little more complicated and deals with the state of a system of atoms and molecules at an absolute zero temperature. Absolute zero is theoretically impossible to achieve considering any force acting upon the atoms and molecules to remove heat from them are not at absolute zero and therefore cannot make anything else reach absolute zero. The third law says that entropy of atoms and molecules at absolute zero is zero.

Environmental Effects of Fossil Fuels

Fossil fuels have many enviromental effects. In fact most big places such as in the The United States, more than 90% of greenhouse gas emissions come from the combustion of fossil fuels.

The combustion of Fossil Feuls can cause acid rain which impact both natural and man-made enviroments. Alsodrilling for fossil fuels off-shore can affect many marine wildlife.

Enviromental Regulation uses a variety of approaches to limit these emissions, such as command and control (which mandates the amount of pollution or the technology used), economic incentives, or voluntary programs.

Coal, oil, and natural gas are America's primary source of energy. Most American's are generally aware that fossil fuels are harmful but are still not fully aware why. The by-products from burning fossil fuels are very dangerous. These small particles can exist in the air for an indefinite amount of time. The particles can even get into the lungs causing several problems. Over a lifetime of continued exposure, a person's ability to transfer oxygen and get rid of pollutants in their body becomes much harder. Those affected could get fatal asthma attacks and other serious lung conditions. Recently, respiratory diseases were found to have caused several disabilities and illnesses. Because the contamination is growing rapidly, the earlier people become aware of the effects of fossil fuels on the environment the faster we can reduce the harmful effects.

There are many products that result from the reaction of burning coal that impact the environment. Sulfur in the coal reacts with oxygen in the air making sulfur oxides. Nitrogen oxides form during the combustion by oxidation of nitrogen in the air. Carbon dioxide is formed by the oxidation of the carbon in the coal. Ozone is made as a product of the reaction of nitric oxide with oxygen. Surface mining also affects the land by damaging surface vegetation. This temporary time can permanently change the landscape as with mountain-top removal and valley fill operations. This is when valleys and streams are filled with the debris from mining. This greatly affects wildlife and their homes. This process changes the water underneath the surface and can ruin drinking water in wells. Coal mining contributes to water pollution. Coal contains pyrite, a sulfur compound and as water washes through the mines, forming a mild acid, which is taken to nearby rivers and streams.

Compared to other fossil fuels, the burning of natural gas is the cleanest. Coal and oil have higher carbon ratios and higher nitrogen and sulfur contents. This means that coal and oil give more harmful emissions when burned. Coal and fuel oil also release ash particles into the environment, the particles not burned are carried into the atmosphere causing pollution. The burning of natural gas releases very small amounts of sulfur dioxide and nitrogen oxides, barely any ash, and lower levels of carbon dioxide and carbon monoxide. The leading pollutant released when burning natural gas is carbon dioxide, but compared to other fossil fuels it is considered very little. The transportation of the gas through pipelines causes most of the pollution from natural gas. Compression stations disrupt the natural wildlife around them. Though natural gas is the cleanest fossil fuel it has a big impact on the ozone layer. Natural gas is mainly made of methane, a greenhouse gas. Methane traps more heat that Carbon Dioxide, which makes people begin to wonder how big of an impact natural gas has on the ozone layer. All in all, natural gas causes less pollution than coal or oil and for right now is the cleanest fossil fuel we have.

The drilling, transportation, and use of oil cause pollution. Oil spills leave waterways and their shores not fit to live in for long periods of time. Spills usually end up losing a majority of plant and animal life in the area. Oil greatly effects wildlife when spills occur. They harm marine birds, mammals, and fish. Many animals even try to clean themselves but are poisoned after ingesting the oil. Oil spills endanger many species. A lot of oil pollution is caused by domestic incidents. Most oil poured down drains from homes finds its way to streams, rivers, and the sea. The largest contribution of oil into the world's oceans is from pouring oil down the drain. There are many incidents when oil is spilled, during delivery or when storage tanks are filled, when storage tanks leak because they are not maintained, or because they are not protected from vandalism. Oil can affect our drinking water when carelessly disposed of in drains or soaked in the ground.

If these issues are not made aware to the American public then the problem will continue to exist in full effect. The best thing to do at this point is make people aware and hopefully people will individually do their part to reduce fossil fuel pollution.

Fossil Fuels such as coal and gasoline provide most of the energy needs of the world today, but because of their diminishing reserves, high prices and most importantly, their damaging effect on the environment, alternative sources of energy and environmentally friendly fuels are now being developed.

From the perspective of protecting the environment, alternative fuels and alternative sources of energy usually fall under seven broad headings.

- Biofuels
- Natural Gas
- Wind Energy
- Hydroelectric Power
- Solar Energy
- Hydrogen
- Nuclear Energy

Biofuels : Any kind of fuels made from plants or animals. These include wood, wood chippings, methane from animal excrement or as a result of bacterial action and ethanol from plant materials. Lately it is ethanol that has become synonymous with the term biofuel and is in wide use in combination with gasoline in the transportation industry.

Natural Gas : Although a fossil fuel, Natural Gas is cleaner burning than gasoline, but does produce Carbon Dioxide, the main greenhouse gas. Like gasoline natural gas is a finite source, but unlike it, there is still a very plentiful supply still available. The EIA, in conjunction with the Oil and Gas Journal and World Oil publications, "estimates world proved natural gas reserves to be around 5,210.8 Tcf (Trillion cubic feet)".

Wind Energy : One of the oldest and cleanest forms of energy and the most developed of the renewable energy sources. There is the potential for a large amount of energy to be produced from wind. The Global Wind Energy Council is forecasting that "the global wind market will grow by over 155% to reach 240 GW of total installed capacity by 2012." Unfortunately wind farms, whether onshore or off shores are unsightly, noisy and generate a lot of opposition.

Hydroelectric Power : Like wind energy, a very old and well developed energy source, but unlike wind energy its capacity for expansion is limited. Over development and unrestricted harnessing of water power can have devastating effect on the local environment and habitation areas.

Solar Energy : Apart from Nuclear Energy, all other forms of energy result from solar energy. Fossil fuels, biofuels and natural gas are in effect "bottled" solar energy. The wind and rivers which provide renewable energy are the result of solar energy reacting with the earth's atmosphere. It is also possible to harness this inexhaustible supply of energy directly through photoelectric cells or using Thermal Power plants.

Hydrogen : Hydrogen could be a very environmentally friendly fuel, and with the advent of the fuel cell it has been proved a viable fuel source for vehicles. But there are serious questions on its production, storage and distribution. There are also questions on its energy efficiency, as so far, it takes more energy to manufacture than it produces.

Nuclear Energy : Once thought to be the "Jewel in the Crown" of alternatives to fossil fuels, Nuclear Energy has received a very bad press since the "Three Mile Island" incident and especially since the Chernobyl accident. Nuclear fission is now a mature and very well understood source of energy, but generates a lot of opposition because of safety concerns. It is very costly and produces difficult to handle toxic waste. Nuclear fusion, which would have no such safety or waste problems, remains the "Holy Grail" of alternative energy, but so far science has failed to come up with a working solution.

No one alternative source will solve the problems posed by global warming. Wind energy does have potential, biofuels and hydrogen are possibilities, but all these have associated problems as well. Coupled with more investment and better technology, the solution should come from a combination of all these sources.

MANAGEMENT OF ENERGY RESOURCES

Since the cost of energy has become a significant factor in the performance of economy of societies, management of energy resources has become very crucial. Energy management involves utilizing the available energy resources more effectively that is with minimum incremental costs. Many times it is possible to save expenditure on energy without incorporating fresh technology by simple management techniques. Most often energy management is the practice of using energy more efficiently by eliminating energy wastage or to balance justifiable energy demand with appropriate energy supply. The process couples energy awareness with energy conservation.

About the management of energy resources

Since the beginning of mankind, energy has been utilized to meet human necessities, which can be classified in four basic energy services: heat, mechanical power, lighting, and communication. At the first stages, energy sources use was a spontaneous process, conditioned to the discovering and developing of new fuels and technologies. During the human society development, the economic criteria became in the determining factor in energy resource allocation problems, specifically the internal costs of energy production processes, which determine the market prices. Nevertheless, these costs only reflect the relations and interests in the sphere of producers and consumers, and not those of all society.

Since the 70's, when the environmental negative impacts became quite evident, some authors have pointed out the necessity of the implementation of multi–criteria analysis in energy planning. This means that other aspects have to be considered together with the economic factor, like environmental and social impacts. In other cases the external costs of energy processes, also called environmental and/or social costs, are incorporated to the energy planning process and reflected in some way in market prices. This latest approach is frequently referred to as "internalization of externalities". At the present time, the methods used to select energy alternatives are linked with interests of social groups addressing the economic policy, and they reflect the governmental position of each country. Regarding rural areas, in the last decades, the main objectives for many policy makers in developing countries were the increasing of agricultural production, education and health, without taking into account the evident negative impacts of energy technologies on environment like pollution by fertilizers and pesticides, deforestation and desertification. On the other hand, they generally have considered the supply side characteristics, but have forgotten the demand side ones, causing the failure of the introducing of new technologies in the rural sector. In this sense, some authors have pointed out that the essence of success in energy planning in rural areas is based on doing an iterative process of matching energy demand and supply. The selection of energy alternatives is a complex process, subordinated to the application context and the interests involved in it.The Management of Energy Resources in Rural Areas software is a tool for evaluation and selection of technological alternatives to satisfy the domestic energy necessities in rural communities, mainly in developing countries.

7

FRESH WATER RESOURCES

Water resources are sources of water that are useful or potentially useful to humans. Uses of water include agricultural, industrial, household, recreational and environmental activities. Virtually all of these human uses require fresh water. 97.5% of water on the Earth is salt water, leaving only 2.5% as fresh water of which over two thirds is frozen in glaciers and polar ice caps. The remaining unfrozen freshwater is mainly found as groundwater, with only a small fraction present above ground or in the air. Fresh water is a renewable resource, yet the world's supply of clean, fresh water is steadily decreasing. Water demand already exceeds supply in many parts of the world, and as world population continues to rise at an unprecedented rate, many more areas are expected to experience this imbalance in the near future. The framework for allocating water resources to water users (where such a framework exists) is known as **water rights.**

Water Right in water law refers to the right of a user to use water from a water source, e.g., a river, stream, pond or source of groundwater. In areas with plentiful water and few users, such systems are generally not complicated or contentious. In other areas, especially arid areas where irrigation is practiced, such systems are often the source of conflict, both legal and physical. Some systems treat surface water and ground water in the same manner, while others use different principles for each.

Overall in the country, the total amount of surface water available is enormous as evident from the annual discharge of many rivers that flow in various parts. There are rivers that flow throughout the year and some flow for limited periods. Generally, the peninsular river flow fully during the monsoon periods and partly in the non-monsoon times; they may even trickle at places due to limited groundwater discharge into the river during such dry periods. In Large Himalayan rivers such as the Ganges, Brahmaputra, Indus, Yamuna and Sutluj there are always water flows though during non-monsoon periods, the flow may be reduced. Most of the rivers in India empty their water into either the Arabian Sea or the Bay of Bengal. Though many of the rivers have dams and diversion canals, potential for water withdrawal from rivers for variety of uses has not yet been fully achieved. Water scarcity is generally seen in those regions of the country where there are no large rivers, diversion canals or storage facility or intensive rainfall.

River basins representing the main source of fresh water in the sub-continent are distributed throughout the country, but many of the minor desert rivers such as Luni and Mahi in Rajasthan and Gujarat are not listed since they flow for a vary duration even during monsoon. Peninsular rivers such as Kaveri do not carry sufficient water in non-monsoon months. In fact, for many such rivers, more than 80% of their annual water flow is carried in the four-month monsoon period. Godavari is as large as the Yamuna but due to glacier melts the later carry water even in summer, though in reduced quantity. So size of a river basin is no indication of its carrying capacity. Narmada and Pranhita (a tributary of Godavari) are about the same size and have about the same carrying capacity though both are non-Himalayan rivers.

The primary sources of water for all the rivers are either dry (snow) or wet (rain) precipitation. Source for both rain and snow has its origin in ocean-atmosphere interaction. Hence the west flowing rivers though small in size, has very high flow rate, expressed in centimeter, due to higher precipitation rates in that part of the region. The flow rate (specific given) of a small river, such as Manimala, with 1% of basin area of that of the Ganges, is thrice that of the Ganges.

Thus, the importance of a river does not lie in its size but how much water can carry per unit area of that river. On the other-hand, Himalayan river Indus is about the same size as that of Godavari but cause much less waters. Since rivers carry dissolved load, the fate of the solute load depends on a number of natural and anthropogenic factors and each component of the solute load will react in different way to changes in external environment. For each type of major processes, example is given for a parameter that is governed by that process such as chloride levels that is generally constant in fresh waters. Whereas the supply rates of mercury (Hg) from land is more than the removal rates leading to Hg accumulation in water column.

At present, more than 85% of the water demand is in the agricultural sector. This is expected to decrease progressively to about 74% by the year 2050; still, the bulk use will be in the irrigation related activities.

While the demand for water in the energy sector is likely to dramatically increase by nearly seventy folds over a 50 years period, demand in the industrial sector is expected to increase eight folds in the same period. The domestic demands, primarily in response to increase in population, will also be increasing. Thus, the total demand on water in the next fifty years is expected to more than double and we will have to find additional source of water for this purpose.

Already our supply rates are far below the demand growth rates; unless we find new resources the per capita availability will drop sharply farther in the years to come. In Australia, in the semi-arid regions, they have just now discovered (News item, Times of India, July 17, 2000) a huge sub-surface reservoir that can meet the demands of the nearby township Perth for over 4000 years. We need to be likewise lucky in our exploration methods for water.

In addition to the rivers, the entire sub-continent is covered with a large number of water bodies reservoirs, lakes, wetlands, mangroves and ponds. Even though the distribution of various types of fresh water bodies are widely distributed in the country, still the availability of drinking water suggests skewed distribution of actual availability. These water bodies are the lungs of the country that regulate both the quantity and quality of water in addition to supporting the biota of various species.

Unfortunately, many such water bodies are either disappearing or shrinking (example many lakes in Rajasthan) or simply getting converted to wastelands due to misuse. During the current water crisis, the largest lake in Udaipur was desilted after 300 years only to find some remains of old and abandoned temple architecture. Similarly, the water levels in the Krishnaraja Sagar Reservoir (in Mettur, Tamilnadu) on the river Kaveri went down very low (exposing some temples) due to lack of water input from the upstream region.

The importance of these water bodies can be seen from the fact that of the thirteen states where there frequent floods and drought in the last few years, nearly 50% of the areas of the states are prone for periodic drought indicating to the vanishing water bodies in those places.

Since most of the rivers end up in oceans, the coastal region of India, about 7515 km in total length, faces a variety of problems due to the contaminated discharges of the rivers flowing through a number of densely populated regions of the country. Many of our coastal areas have problems connected with water quality due to urban waste discharge. Actual discharge of effluents along the coast from hinterland region and the coastal region of Andhra Pradesh accounts for bulk of the discharge leading to many harmful effects for a variety of health related problems all along the.

Obviously, water quality is as important as the quantity of water that was discussed at length so far. Let us look into key elements of water quality that make up the overall suitability of available water really usable for one purpose or the other.

Uses of fresh water

Uses of fresh water can be categorized as consumptive and non-consumptive (sometimes called "renewable"). A use of water is consumptive if that water is not immediately available for another use. Losses to sub-surface water and evaporation are considered consumptive, as is water incorporated into a product (such as farm produce). Water that can be treated and returned as surface water, such as sewage, is generally considered non-consumptive if that water can be put to additional use.

Sources of fresh water–

Surface water : Surface water is water in a river, lake or fresh water wetland. Surface water is naturally replenished by precipitation and naturally lost through discharge to the oceans, evaporation, and sub-surface water.

Although the only natural input to any surface water system is precipitation within its watershed, the total quantity of water in that system at any given time is also dependent on many other factors. These factors include storage capacity in lakes, wetlands and artificial reservoirs, the permeability of the soil beneath these storage bodies, the runoff characteristics of the land in the watershed, the timing of the precipitation and local evaporation rates. All of these factors also affect the proportions of water lost.

Human activities can have a large impact on these factors. Humans often increase storage capacity by constructing reservoirs and decrease it by draining wetlands. Humans often increase runoff quantities and velocities by paving areas and channelizing stream flow.

The total quantity of water available at any given time is an important consideration. Some human water users have an intermittent need for water. For example, many farms require large quantities of water in the spring, and no water at all in the winter. To supply such a farm with water, a surface water system may require a large storage capacity to collect water throughout the year and release it in a short period of time. Other users have a continuous need for water, such as a power plant that requires water for cooling. To supply such a power plant with water, a surface water system only needs enough storage capacity to fill in when average stream flow is below the power plant's need.

Nevertheless, over the long term the average rate of precipitation within a watershed is the upper bound for average consumption of natural surface water from that watershed.

Natural surface water can be augmented by importing surface water from another watershed through a canal or pipeline. It can also be artificially augmented from any of the other sources listed here, however in practice the quantities are negligible. Humans can also cause surface water to be "lost" (i.e. become unusable) through pollution.

Brazil is the country estimated to have the largest supply of fresh water in the world, followed by Russia and Canada.

Sub-surface water : Sub-surface water, or groundwater, is fresh water located in the pore space of soil and rocks. It is also water that is flowing within aquifers below the water table. Sometimes it is useful to make a distinction between sub-surface water that is closely associated with surface water and deep sub-surface water in an aquifer (sometimes called "fossil water").

Sub-surface water can be thought of in the same terms as surface water: inputs, outputs and storage. The critical difference is that due to its slow rate of turnover, sub-surface water storage is generally much larger compared to inputs than it is for surface water. This difference makes it easy for humans to use sub-surface water unsustainably for a long time without severe consequences. Nevertheless, over the long term the average rate of seepage above a sub-surface water source is the upper bound for average consumption of water from that source.

The natural outputs from sub-surface water are springs and the oceans.

If the surface water source is also subject to substantial evaporation, a sub-surface water source may become saline. This situation can occur naturally under endorheic bodies of water, or artificially under irrigated farmland. In coastal areas, human use of a sub-surface water source may cause the direction of water to ocean to reverse which can also cause soil salinization. Humans can also cause sub-surface water to be "lost" (i.e. become unusable) through pollution. Humans can increase the input to a sub-surface water source by building reservoirs or detention ponds.

Water in the ground is in sections called aquifers. Rain rolls down and comes into these. Normally an aquifer is near the equilibrium in its water content. The water content of an aquifer normally depends on the grain sizes. This means that the rate of extraction may be limited by poor permeability.

Desalination : Desalination is an artificial process by which saline water (generally sea water) is converted to fresh water. The most common desalination processes are distillation and reverse osmosis. Desalination is currently expensive compared to most alternative sources of water, and only a very small fraction of total human use is satisfied by desalination. It is only economically practical for high-valued uses (such as household and industrial uses) in arid areas. The most extensive use is in the Persian Gulf.

Frozen water : Several schemes have been proposed to make use of icebergs as a water source, however to date this has only been done for novelty purposes. Glacier runoff is considered to be surface water.

Ground Water : The importance of groundwater for the existence of human society cannot be overemphasized. Groundwater is the major source of drinking water in both urban and rural India. Besides, it is an important source of water for the agricultural and the industrial sector. Water utilization projections for 2000 put the groundwater usage at about 50%. Being an important and integral part of the hydrological cycle, its availability depends on the rainfall and recharge conditions. Till recently it had been considered a dependable source of uncontaminated water.

The demand for water has increased over the years and this has led to water scarcity in many parts of the world. The situation is aggravated by the problem of water pollution or contamination. India is heading towards a freshwater crisis mainly due to improper management of water resources and environmental degradation, which has lead to a lack of access to safe water supply to millions of people. This freshwater crisis is already evident in many parts of India, varying in scale and intensity depending mainly on the time of the year.

Groundwater crisis is not the result of natural factors; it has been caused by human actions. During the past two decades, the water level in several parts of the country has been falling rapidly due to an increase in extraction. The number of wells drilled for irrigation of both food and cash crops have rapidly and indiscriminately increased. India's rapidly rising population and changing lifestyles has also increased the domestic need for water. The water requirement for the industry also shows an overall increase.

As far as the quality of groundwater is concerned, many states in the country have been identified as endemic to fluorosis due to abundance in naturally occurring fluoride bearing minerals. These are Andhra Pradesh, Gujarat, Haryana, Orissa, Punjab, Rajasthan, Tamil Nadu, Uttar Pradesh, Karnataka, Madhya Pradesh, Maharastra, Bihar, and Delhi. Nearly half million people in India suffer from ailments due to excess of fluoride in drinking water. In some districts of Assam and Orissa, groundwater has high iron content. About 31% of the total area of Rajasthan comes under saline groundwater. Groundwater is saline in almost all of the Bhakra Canal in Punjab and the lift canal system in south-western Haryana. Similarly high levels of arsenic in groundwater have been reported in the shallow aquifers in some districts of West Bengal. Certain places in Haryana, Gujarat, and Andhra Pradesh were also found to have dangerously high levels of mercury.

Flood

A **flood** is an overflow of an expanse of water that submerges land, a deluge. In the sense of "flowing water", the word may also be applied to the inflow of the tide. Flooding may result from the volume of water within a body of water, such as a river or lake, which overflows, with the result that some of the water escapes its normal boundaries. While the size of a lake or other body of water will vary with seasonal changes in precipitation and snow melt, it is not a significant flood unless such escapes of water endangers land areas used by man like a village, city or other inhabited area.

Floods can also occur in rivers, when the strength of the river is so high it flows out of the river channel, particularly at bends or meanders and cause damage to homes and businesses along such rivers. While flood damage can be virtually eliminated by moving away from rivers and other bodies of water, since time out of mind, man has lived and worked by the water to seek sustenance and capitalize on the gains of cheap and easy travel and commerce by being near water. That humans continue to inhabit areas threatened by flood damage is only evidence that the value of being near the water far exceeds the costs of repeated periodic flooding.

Causes of Floods

Many floods are directly related to changes in **weather**. The most common cause of flooding is due to rain falling at extremely high rates or for an unusually long period of time. Additionally, areas that experience a great deal of snow in winter are prone to springtime flooding when the snow and ice melt, especially if the thaw is relatively sudden. Furthermore, rainfall and snowmelt can sometimes combine to cause floods.

Floods are caused by a variety of factors, both natural and man-made. Some obvious causes of floods are heavy rains, melting snow and **ice**, and frequent storms within a short time duration. The common practice of humans to build homes and towns near rivers and other bodies of water (i.e., within natural floodplains) has contributed to the disastrous consequences of floods. In fact, floods have historically killed more people than any other form of natural disaster. Because of this fact, humans have attempted to manage floods using a variety of methods with varying degrees of success.

Sometimes, floods occur as a result of a unique combination of factors that only indirectly involve weather conditions. For instance, a low-lying coastal area may be prone to flooding whenever the **ocean** is at high tide. Exceptionally high **tides** may be attributed to a **storm** caused by a combination of factors, like low barometric **pressure** and high winds. Finally, floods sometimes can occur regardless of the climate. Examples are tsunamis (seismic waves on the sea or large lakes that are caused by earthquakes), volcanic heating and rapid melting of a snow pack at top of a volcanic mountain or under a glacier, or even failures of natural or man-made **dams**.

Effects Of Floods

Primary effects :

- *Physical damage* - Can range anywhere from bridges, cars, buildings, sewer systems, roadways, canals and any other type of structure.

- *Casualties* - People and livestock die due to drowning. It can also lead to epidemics and diseases.

Secondary effects :

- *Water supplies* - Contamination of water. Clean drinking water becomes scarce.
- *Diseases* - Unhygienic conditions. Spread of water-borne diseases.
- *Crops and food supplies* - Shortage of food crops can be caused due to loss of entire harvest. However, lowlands near rivers depend upon river silt deposited by floods in order to add nutrients to the local soil.
- *Trees* - Non-tolerant species can die from suffocation.

Tertiary/long-term effects :

- *Economic* - Economic hardship, due to: temporary decline in tourism, rebuilding costs, food shortage leading to price increase etc.

Flooding produces many effects in the region. It not only damages property and endangers the lives of humans or animals, but can have other effects as well. Rapid water runoff causes soil erosion in addition to sediment deposition problems further downstream. The spawning grounds for fish and other wildlife habitats can become polluted or completely destroyed. Some prolonged high floods can delay traffic in areas which lack elevated roadways. Floods can interfere with drainage and economic use of lands, such as interfering with farming. Structural damage can occur in bridge abutments, bank lines, sewer lines, and other structures within floodways. Waterway navigation and hydroelectric power are often impaired. Financial losses due to floods are typically millions of dollars each year.

Methods of control

In many countries, rivers prone to floods are often carefully managed. Defences such as levees, bunds, reservoirs, and weirs are used to prevent rivers from bursting their banks. Coastal flooding has been addressed in Europe and the Americas with coastal defences, such as sea walls, beach nourishment, and barrier islands.

Dams : A **dam** is a barrier that impounds water or underground streams. Dams generally serve the primary purpose of retaining water, while other structures such as floodgates, levees, and dikes are used to manage or prevent water flow into specific land regions.

Ground Water Dams : Groundwater dams are structures that intercept or obstruct the natural flow of groundwater and provide storage for water underground. They have been used in several parts of the world, notably India, Africa and Brazil. Their use is in areas where flows of groundwater vary considerably during the course of the year, from very high flows following rain to negligible flows during the dry season.

The basic principle of the groundwater dam is that instead of storing the water in surface reservoirs, water is stored underground. The main advantages of water storage in groundwater dams is that evaporation losses are much less for water stored underground. Further, risk of contamination of the stored water from the surface is reduced because as parasites cannot breed in underground water. The problem of submergence of land which is normally associated with surface dams is not present with sub-surface dams.

There are two main types of groundwater dam: the sub-surface dam and the sand storage dam.

A sub-surface dam intercepts or obstructs the flow of an aquifer and reduces the variation of the level of the groundwater table upstream of the dam. It is built entirely under the ground (see figure 1).

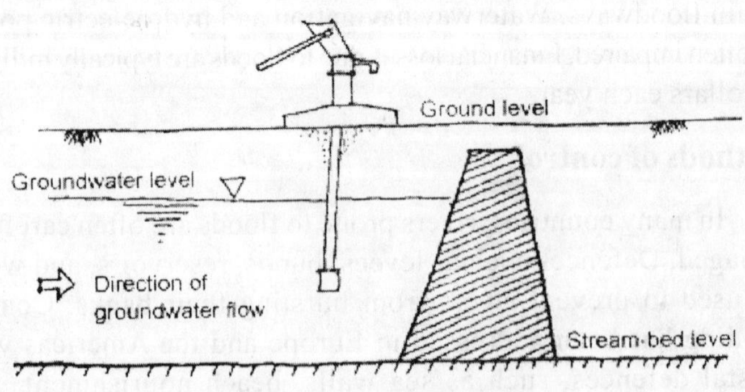

Fig. 1 : A sub-surface dam

The sand storage dam is constructed above ground. Sand and soil particles transported during periods of high flow are allowed to deposit behind the dam, and water is stored in these soil deposits (see figure 2). The sand storage dam is constructed in layers to allow sand to be deposited and finer material be washed downstream.

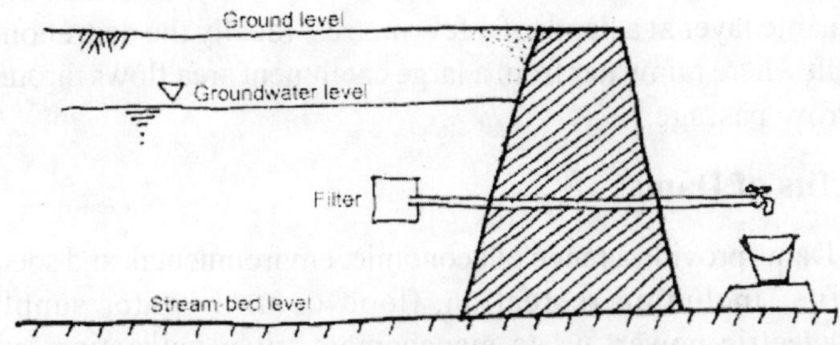

Fig. 2 : A sand storage dam

A groundwater dam can also be a combination of these two types. When constructing a sub-surface dam in a river bed, one can increase the storage volume by letting the dam wall rise over the surface, thus causing additional accumulation of sediments. Similarly, when a sand-storage dam is constructed it is necessary to excavate a trench in the sand bed in order to reach bedrock, which can be used to create a sub-surface dam too.

Groundwater dams are built across streams or valleys. A trench is dug across the valley or stream, reaching to the bedrock or other stable layer like clay. An impervious wall is constructed in the trench, which is then refilled with the excavated material.

Various materials may be used for the construction of groundwater dams. Materials should be waterproof, and the dam should be strong enough to withstand the imposed soil and water loads. Dams may vary from 2 to 10 metres high. Materials include compacted clay, concrete, stones and clay, masonry wall or plastic sheets.

Stages of construction of a sand-storage dam : The reservoir is recharged during the monsoon period and the stored water can be used during the dry season. Excess water flows over the top of the dam to replenish aquifers downstream. Water may be obtained from the underground reservoir either from a well upstream of the dam or from a pipe, passing through the dam, and leading to a collection point downstream. Groundwater dams cannot be a universally applicable as these require specific conditions for functioning. The best sites for construction of groundwater dams are where the soil consists of sands and gravel, with rock or a permeable layer at a depth of a few metres. Ideally the dam should be built where rainwater from a large catchment area flows through a narrow passage.

Benefits of Dams

Dams provide a range of economic, environmental, and social benefits, including recreation, flood control, water supply, hydroelectric power, waste management, river navigation, and wildlife habitat.

- **Recreation** : Dams provide prime recreational facilities throughout the United States. Boating, skiing, camping, picnic areas, and boat launch facilities are all supported by dams.
- **Flood Control :** In addition to helping farmers, dams help prevent the loss of life and property caused by flooding. Flood control dams impound floodwaters and then either release them under control to the river below the dam or store or divert the water for other uses. For centuries, people have built dams to help control devastating floods.
- **Water Storage :** Dams create reservoirs throughout the United States that supply water for many uses, including industrial, municipal, and agricultural.
- **Irrigation :** Ten percent of American cropland is irrigated using water stored behind dams. Thousands of jobs are tied to producing crops grown with irrigated water.
- **Mine Tailings :** There are more than 1,300 mine tailings impoundments in the United States that allow the mining and processing of coal and other vital minerals while protecting the environment.
- **Electrical Generation :** The United States is one of the largest producers of hydropower in the world, second only to

Canada. Dams produce over 103,800 megawatts of renewable electricity and meet 8 to 12 percent of the Nation's power needs. Hydropower is considered clean because it does not contribute to global warming, air pollution, acid rain, or ozone depletion.

• **Debris Control :** In some instances, dams provide enhanced environmental protection, such as the retention of hazardous materials and detrimental sedimentation.

• **Navigation :** Dams and locks provide for a stable system of inland river transportation throughout the heartland of the Nation.

The problems with dams

What's wrong with dams?

Dams kill rivers.

Dams can have devastating effects on rivers and freshwater ecosystems. Only very few of the world's rivers now run uninterrupted from their source to the sea.

Fragmentation of rivers affects the migration of fish, disrupts the transport of sediments, cuts off the floodplains from life giving floods and threatens many endangered species, including the last populations of river dolphins. Environmental problems extend throughout the river basin.

Destroy livelihoods : The devastation of freshwater ecosystems directly affects the livelihoods of millions of people who live downstream of dams, especially in developing countries. And this is in addition to the large numbers of people that have to make way for the reservoirs.

And cost a great deal of money : finally, dams are expensive and the projected financial costs are often inaccurate. The World Commission on Dams found that, on average, large dams go over budget by 56%. And since these days most of the world's dams are being built in developing countries, it is even more important that a project makes economic sense. Countries in poverty simply cannot afford to pay for mistakes.

Drought

A **drought** is an extended period of months or years when a region notes a deficiency in its water supply. Generally, this occurs when a region receives consistently below average precipitation. It can have a substantial impact on the ecosystem and agriculture of the affected region. Although droughts can persist for several years, even a short, intense drought can cause significant damage and harm the local economy. This global phenomenon has a widespread impact on agriculture. The United Nations estimates that an area of fertile soil the size of Ukraine is lost every year because of drought, deforestation, and climate instability. Lengthy periods of drought have triggered mass migration in Africa in this last decade and in various other parts of the world for thousands of years.

Drought is a temporary reduction in water or moisture availability significantly below the normal or expected amount for a specific period. This condition occurs either due to inadequacy of rainfall, or lack or irrigation facilities, under-exploitation or deficient availability for meeting the normal crop requirements in the context of the agro-climatic conditions prevailing in any particular area. This has been scientifically computed as Moisture index (M I). There is a drought in jaisalmer (Average rainfall 200 mm) if rainfall is not sufficient to grow grass an paltry coarse-grains, whereas in Bolangir or Koraput (Orissa-rainfall above 1000 mm) there is a drought if there is not enough rainfall for bringing the paddy crop to maturity.

Droughts can be of three kinds

(i) **Meteorological drought**: This happens when the actual rainfall in an area is significantly less than the climatological mean of that area. The country as a whole may have a normal monsoon, but different meteorological districts and sub-divisions can have below normal rainfall. The rainfall categories for smaller areas are defined by their deviation from a meteorological area's normal rainfall -

Excess: 20 per cent or more above normal

Normal: 19 per cent above normal - 19 per cent below normal

Deficient: 20 per cent below normal - 59 per cent below normal

Scanty: 60 per cent or more below normal

(ii) **Hydrological drought**: A marked depletion of surface water causing very low stream flow and drying of lakes, rivers and reservoirs

(iii) **Agricultural drought**: Inadequate soil moisture resulting in acute crop stress and fall in agricultural productivity.

Biota

Biota is the total collection of organisms of a geographic region or a time period, from local geographic scales and instantaneous temporal scales all the way up to whole-planet and whole-timescale spatiotemporal scales. The biota of the Earth lives in the biosphere.

Conservation of Fresh Water

Water conservation refers to reducing the use of water.

The goals of water conservation efforts include:

- **Sustainability** :To ensure availability for future generations, the withdrawal of fresh water from an ecosystem should not exceed its natural replacement rate.
- **Energy conservation** : Water pumping, delivery, and wastewater treatment facilities consume a significant amount of energy. In some regions of the world (for example, California).
- **Habitat conservation** : Minimizing human water use helps to preserve fresh water habitats for local wildlife and migrating waterfowl, as well as reducing the need to build new dams and other water diversion infrastructure.

Over 99% of biosphere water occurs in oceans and polar ice deposits, out of which 97.61% occurs in oceans. The freshwater is mainly in the form of ice, snow and ground water, 0.009% in freshwater lakes, 0.0009% in atmospheric water vapour and 0.00009% in rivers. Only 0.01% of the global freshwater is available in rivers, lakes and reservoirs. The dam reservoirs contain five times as much water as in rivers. The surface waters sustain freshwater biodiversity, perform ecological functions and support human needs such as agriculture, hydro-electricity, industry, sewage and sanitation, aquaculture, fisheries, drinking water, transportation, recreation and spiritual needs, etc. About 54% of accessible surface run off is used. About 45,000 species of freshwater organisms are known while about one million are yet to be discovered. Major organisms include viruses, bacteria, diatoms, plants and animals from protozoa to mammals. Freshwater organisms constitute about 25% of the total number of organisms.

Aquatic bacterial diversity knowledge is increasing rapidly. Almost 20% of the fishes, found globally, are extinct, endangered. Groundwaters, as deep as 2.8 Km, may have rich bacterial flora. According to World Bank, 80 countries with 40% of the world's population have water shortage that could cripple agriculture and industry. Fish diversity is threatened by construction of dams, hydro-electric facilities, channelization projects and invasion of non-native biota. The Ganges and Brahmaputra rivers carry more than 3 billion metric tons of soil to the Bay of Bengal each year, spreading it over 3 million sq. km of sea bed.

Benefits from Freshwater Biodiversity

Freshwater biodiversity provides benefits to humans. This includes inland water fishing for food, aquaculture production, ornamental fish trade, recreational fishing, rice farming, harvest of a variety of other living resources, medicinal plants, fuel resources and ecological functions including primary production, provision of three dimensional habitat, biogeochemical recycling, pollutant remediation, moderation of nutrient pulses and population . Terrestrial and aquatic ecological functions have been estimated to be worth US $ 33 trillion per year globally. Freshwaters and their varied biodiversity form part of the Earth's ecosphere which restores us spiritually, inspires us aesthetically and must be passed on to future generations. The Biodiversity Convention and the World's Charter of Nature have emphasized that like all other life forms, freshwater organisms have an intrinsic right to survival and warrant respect.

Reservoir

A **reservoir** is, most broadly, a place or hollow vessel where fluid is kept in reserve, for later use. Most often, a reservoir refers to an artificial lake, used to store water for various uses. Reservoirs are created first by building a sturdy dam, usually out of cement, earth, rock, or a mixture. Once the dam is completed, a stream is allowed to flow behind it and eventually fill it to capacity. When such a reservoir is predominantly man-made (rather than being an adaptation of a natural structure) it may be called a cistern. The term reservoir is also often used to describe underground reservoirs such as an oil or water well.

There are two basic types of reservoir: the commonly-seen dam across a valley, or the less-common fully-bunded dam.

A *fully water tower bunded dam* has a continuous human made embankment around its entire perimeter, most commonly using a central clay core as the waterproof element. The core is held in place by earth or rock piled either side of it in suitable volumes to resist the outward forces exerted by the water. The clay is joined directly to the natural underlying material, which itself is usually clay in order for the dam to be watertight. The reservoir is filled by mechanical pumps that draw water from an adjacent water course such as a river. If a watertight roof is added then it can then be used for storing treated water before it goes to the tap. This is known as a "service" reservoir.

Operation

A raw water reservoir doesn't simply hold water until it is needed. It is the first part of the water treatment process. The time the water is held for before it is released is known as the *retention time*. This is a design feature that allows particles and silts to settle out, as well as time for the biological treatment of algae and bacteria by plankton like creatures that naturally live within the water.

Water can be released from the reservoir, generally by gravity, to be cleaned for drinking water, generate electricity, or simply maintain the downstream flow. In the event that a major rain occurs, water can be released via a spillway to avoid overtopping and compromising the integrity of the dam.

8

FOOD RESOURCES

Global food production, so far, has increased continuously because cropped area has expanded and productivity per unit area has increased. In some regions of the world, however, there is little scope for further spatial expansion of agriculture.

While there is an upper limit to food production, results of recent analysis indicate that global agriculture is still far from it. Because decreasing growth in global food production has been observed, policymakers should not be misled into thinking that the world is approaching that limit.

In this research nine food-demand scenarios were analyzed, ranging from minimum population growth combined with a vegetarian diet to maximum population growth combined with an affluent diet containing an ample amount of animal products. Food demand for the intermediate scenario (medium population growth with a moderate diet) is compared with the potential levels of food production

A region with a ratio of 1.0 or less cannot match food consumption with production.

At a global level, four times more food can be produced than required using environment-oriented agriculture and nine times more using ecotechnology-oriented agriculture. When ecotechnology-oriented agriculture is practiced in the reference demand scenario, all regions can provide all of the food necessary. However, using environment-oriented agriculture, some regions in Asia cannot produce enough food to meet their needs or can produce barely enough, even with maximum utilization of natural resources.

Ratios for the extreme scenarios (maximum population with an affluent diet and minimum population with a vegetarian diet) are shown by line marks at the top of the bars in the figure. The ratio is almost twice as high for the minimum-demand scenario and half for the maximum- demand scenario.

With an environment-oriented agriculture, all regions can produce the food required even for an affluent diet, except for East, South, and Southeast Asia; the three regions with the least leeway will carry almost half of the global population. West Asia and West and North Africa come close to the lower limit. A much less expensive diet provides the only option for escape, apart from massive food imports. Europe, the former U.S.S.R., North America, Oceania, South America, and Central Africa are well-off and need only part of their suitable land to feed their populations whatever their diet. However, if trade can distribute food efficiently across the globe, all people may consume an affluent diet, but at the expense of intensive use of two-thirds of the globe for arable crops and rangeland.

In all cases, three times more land is required for environment-oriented agricultural production systems than for ecotechnology-oriented systems. Consequently, the choice of the production technique has a major effect on global land use. Depending on the diet selected, Europe can grow an adequate food supply on 30-50 percent of its suitable land, North America on 20 percent of its land, and South America and Oceania on even smaller fractions.

The benefits of globalization in food and agriculture could outweigh the risks and costs. For example, globalization has generally led to progress in reducing poverty in Asia. "But it has also led to the rise of multinational food companies with the potential to disempower farmers in many countries. Developing countries need the legal and administrative framework to ward off the threats while reaping the benefits." Openess towards international markets, investments in infrastructure, the promotion of economic integration and limits on market concentration, could make globalization work for the benefit of the poor.

Hunger and the Globalized System of Tradeand Food Production

To solve the world hunger crisis, it's necessary to do more than send emergency food aid to countries facing famine. Leaders must address the globalized system of agricultural production and trade that favors large corporate agriculture and export-oriented crops while discriminating against small-scale farmers and agriculture oriented to local needs. As a result of official inaction, more than thirty million people die of malnutrition and starvation every year, while large industrial farms export ever more strawberries and cut flowers to affluent consumers. Excessive meat production, again largely for the affluent, requires massive amounts of feed grains that might otherwise sustain poor families. Giant agribusiness, chemical and restaurant companies like Cargill, Monsanto and McDonalds dominate the world's food chain, building a global dependence on unhealthy and genetically dangerous products. These companies are racing to secure patents on every plant and living organism and their intensive advertising seeks to persuade the world's consumers to eat more and more sweets, snacks, burgers, and soft drinks.

Global Food Production

Promoting healthy diets and lifestyles to reduce the global burden of non-communicable diseases requires a multisectoral approach involving the various relevant sectors in societies. The agriculture and food sector figures prominently in this enterprise and must be given due importance in any consideration of the promotion of healthy diets for individuals and population groups. Food strategies must not merely be directed at ensuring food security for all, but must also achieve the consumption of adequate quantities of safe and good quality foods that together make up a healthy diet. Any recommendation to that effect will have implications for all components in the food chain. It is therefore useful at this juncture to examine trends in consumption patterns worldwide and deliberate on the potential of the food and agriculture sector to meet the demands and challenges posed by this report.

Introduction

Promoting healthy diets and lifestyles to reduce the global burden of non-communicable diseases requires a multisectoral approach involving the various relevant sectors in societies. The agriculture and food sector figures prominently in this enterprise and must be given due importance in any consideration of the promotion of healthy diets for individuals and population groups. Food strategies must not merely be directed at ensuring food security for all, but must also achieve the consumption of adequate quantities of safe and good quality foods that together make up a healthy diet. Any recommendation to that effect will have implications for all components in the food chain. It is therefore useful at this juncture to examine trends in consumption patterns worldwide and deliberate on the potential of the food and agriculture sector to meet the demands and challenges posed by this report.

Economic development is normally accompanied by improvements in a country's food supply and the gradual elimination of dietary deficiencies, thus improving the overall nutritional status of the country's population. Furthermore, it also brings about qualitative changes in the production, processing, distribution and marketing of food. Increasing urbanization will also have consequences for the dietary patterns and lifestyles of individuals, not all of which are positive. Changes in diets, patterns of work and leisure - often referred to as the "nutrition transition" - are already contributing to the causal factors underlying non-communicable diseases even in the poorest countries. Moreover, the pace of these changes seems to be accelerating, especially in the low-income and middle-income countries.

The dietary changes that characterize the "nutrition transition" include both quantitative and qualitative changes in the diet. The adverse dietary changes include shifts in the structure of the diet towards a higher energy density diet with a greater role for fat and added sugars in foods, greater saturated fat intake (mostly from animal sources), reduced intakes of complex carbohydrates and dietary fibre, and reduced fruit and vegetable intakes (*1*). These dietary changes are compounded by lifestyle changes that reflect reduced physical activity at work and during leisure time (*2*). At the same time, however, poor countries continue to face food shortages and nutrient inadequacies.

Diets evolve over time, being influenced by many factors and complex interactions. Income, prices, individual preferences and beliefs, cultural traditions, as well as geographical, environmental, social and economic factors all interact in a complex manner to shape dietary consumption patterns. Data on the national availability of the main food commodities provide a valuable insight into diets and their evolution over time. FAO produces annual Food Balance Sheets which provide national data on food availability (for almost all commodities and for nearly all countries). Food Balance Sheets give a complete picture of supply (including production, imports, stock changes and exports) and utilization (including final demand in the form of food use and industrial non-food use, intermediate demand such as animal feed and seed use, and waste) by commodity. From these data, the average per capita supply of macronutrients (i.e. energy, protein, fats) can be derived for all food commodities. Although such average per capita supplies are derived from national data, they may not correspond to actual per capita availability, which is determined by many other factors such as inequality in access to food. Likewise, these data refer to "average food available for consumption", which, for a number of reasons (for example, waste at the household level), is not equal to average food intake or average food consumption. In the remainder of this chapter, therefore, the terms "food consumption" or "food intake" should be read as "food available for consumption".

Actual food availability may vary by region, socioeconomic level and season. Certain difficulties are encountered when estimating trade, production and stock changes on an annual scale. Hence three-year averages are calculated in order to reduce errors.

The FAO statistical database (FAOSTAT), being based on national data, does not provide information on the distribution of food within countries, or within communities and households.

Food consumption expressed in kilocalories (kcal) per capita per day is a key variable used for measuring and evaluating the evolution of the global and regional food situation. A more appropriate term for this variable would be "national average apparent food consumption" since the data come from national Food Balance Sheets rather than from food consumption surveys. Analysis of FAOSTAT data shows that dietary energy measured in kcals per capita per day has been steadily increasing on a worldwide basis; availability of calories per capita from the mid-1960s to the late 1990s increased globally by approximately 450 kcal per capita per day and by over 600 kcal per capita per day in developing countries

This change has not, however, been equal across regions. The per capita supply of calories has remained almost stagnant in sub-Saharan Africa and has recently fallen in the countries in economic transition. In contrast, the per capita supply of energy has risen dramatically in East Asia (by almost 1000 kcal per capita per day, mainly in China) and in the Near East/North Africa region (by over 700 kcal per capita per day).

Globally, the share of dietary energy supplied by cereals appears to have remained relatively stable over time, representing about 50% of dietary energy supply. Recently, however, subtle changes appear to be taking place (see Fig. 1). A closer analysis of the dietary energy intake shows a decrease in developing countries, where the share of energy derived from cereals has fallen from 60% to 54% in a period of only 10 years. Much of this downwards trend is attributable to cereals, particularly wheat and rice, becoming less preferred foods in middle-income countries such as Brazil and China, a pattern likely to continue over the next 30 years or so. (Fig.2) shows the structural changes in the diet of developing countries over the past 30-40 years and FAO's projections to the year 2030.

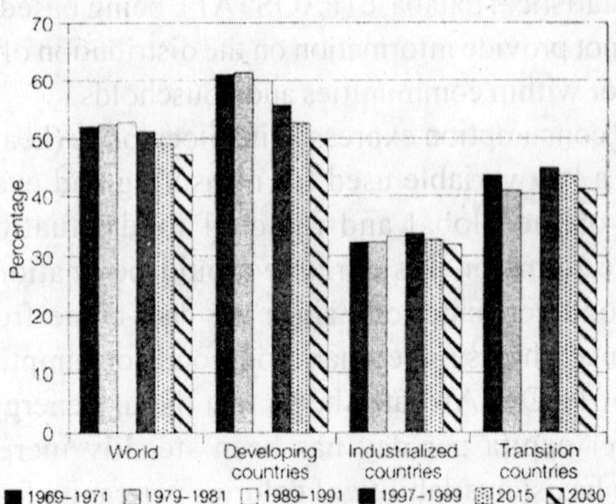

Figure 1. The share of dietary energy derived from cereals

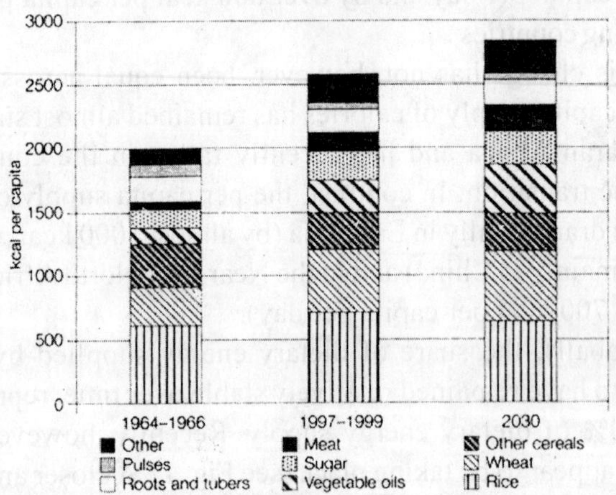

Figure 2. Calories from major commodities in developing countries

Availability and changes in consumption of dietary fat

The increase in the quantity and quality of the fats consumed in the diet is an important feature of nutrition transition reflected in the national diets of countries. There are large variations across the regions of the world in the amount of total fats (i.e. fats in foods, plus added fats and oils) available for human consumption. The lowest quantities consumed are recorded in Africa, while the highest consumption occurs in parts of North America and Europe.

The important point is that there has been a remarkable increase in the intake of dietary fats over the past three decades and that this increase has taken place practically everywhere except in Africa, where consumption levels have stagnated. The per capita supply of fat from animal foods has increased, respectively, by 14 and 4 g per capita in developing and industrialized countries, while there has been a decrease of 9 g per capita in transition countries.

Plant sources of food

Nitrogen : For soils that are nitrogen deficient, treat, manure, weeds, grass clippings, and other garden wastes with CBPA and add the resulting compost to the soil. This will result in a simultaneous improvement in humus and nitrogen content.

Sources of potassium include, dead plant material, manure , compost, granite dust, greensand, and seaweed.

Phosphorus : A natural phosphate rock product can be the best source of phosphorus. Phosphate rock is most effective when applied in combination with manure (3 to 5 lbs. of manure per pound of phosphate). Treat the manure with CBPA and work into the soil. Wait 20 to 30 days and work the phosphate rock into the top layer of soil.

Potassium : Sources of potassium include, dead plant material, manure , compost, granite dust, greensand, and seaweed.

Animal Sources of Food

Animal source foods (ASF) include any food item that comes from an animal source such as meat, milk, fish, eggs, cheese and yogurt. Many individuals do not consume ASF or consume little ASF by either personal choice or necessity as ASF may not be accessible or available to these people

Nutrition of animal source foods

Nutritionally, all nutrients found in animal source foods may be replaced by plant-derived foods. Examples are tofu to replace meat (both contain protein in sufficient amounts), and certain seaweeds and vegetables as respectively kombu and kale to replace dairy foods as milk (both contain calcium in sufficient amounts). Certain plant-derived foods are even nutrient-denser than their animal-derived counterparts (eg tofu).

Although a healthy diet containing all essential macro and micronutrients is possible by only consuming a plant based diet, some populations are unable to consume an adequate quantity or variety of these plant based items to obtain appropriate amounts of nutrients, particularly those that are found in high concentrations in ASF. Frequently, the most vulnerable populations to these micronutrient deficiencies are pregnant women, infants, and children in developing countries. In the 1980s the Nutrition Collaborative Research Support Program (NCRSP) found that six micronutrients were low in the mostly vegetarian diets of children in malnourished areas of Egypt, Mexico, and Kenya. These six micronutrients are vitamin A, vitamin B12, riboflavin, calcium, iron and zinc. ASF are the only food source of Vitamin B12. ASF also provide high biological value protein, energy, fat compared with plant food sources.

Health impacts of micronutrient deficiency

All six micronutrients richly found in ASF, vitamin A, vitamin B12, riboflavin, calcium, iron and zinc play a critical role in the growth and development of children. Inadequate stores of these micronutrients, either resulting from inadequate intake or poor absorption, is associated with poor growth, anemias (iron deficiency anemia and macrocytic anemia), rickets, night blindness, impaired cognitive functioning, neuromuscular deficits, diminished work capacity, psychiatric disorders and death. Some of these affects, such as impaired cognitive development from an iron deficiency, are irreversible even with supplement treatment.

Animal source foods can provide a variety of micronutrients that are difficult to obtain in adequate quantities from plant source foods alone.

In the 1980s, the Nutrition Collaborative Research Support Program identified six micronutrients that were particularly low in the primarily vegetarian diets of schoolchildren in rural Egypt, Kenya and Mexico: vitamin A, vitamin B-12, riboflavin, calcium, iron and zinc. Negative health outcomes associated with inadequate intake of these nutrients include anemia, poor growth, rickets, impaired cognitive performance, blindness, neuromuscular deficits and eventually, death. Animal source foods are particularly rich sources of all six of these nutrients, and relatively small amounts of these foods, added to a vegetarian diet, can substantially increase nutrient adequacy. Snacks designed for children provided more nutrients when animal and plant foods were combined. A snack that provided only 20% of a child's energy requirement could provide 38% of the calcium, 83% of the vitamin B-12 and 82% of the riboflavin requirements if milk was included. A similar snack that included ground beef rather than milk provided 86% of the zinc and 106% of the vitamin B-12 requirements, as well as 26% of the iron requirement. Food guides usually recommend several daily servings from animal source food groups (dairy products and meat or meat alternatives). An index that estimates nutrient adequacy based on adherence to such food guide recommendations may provide a useful method of quickly evaluating dietary quality in both developing and developed countries.

Animal Husbandry in India

Man has been domesticating animals either for cultivation and transport or for food. He has been able to rear good varieties of animals like sheep, birds, cattle,etc.

Importance of Domestic Animals–

- India is primarily an agricultural country. The villagers mainly depend on livestock like cows, buffaloes, goats, fows, horses etc. These animals are grouped into three types depending on their utility.
- Milk-Yielding animals:- The milk-yielding animals are cows, buffaloes, goats, etc. In our country the buffaloes form the major group of milk-yielding cattle, which form the basis of dairy industry. There were about 343 millions of cattle in our country during the year 1965-66. Out of these 176 millions were sheep, 64 millions were goats and the rest were others.
- Animals from which meat is procured and egg laying animals:- Meat for human consumption can be had from sheep, goat, fowls, ducks etc. Out of these the fowls and ducks lay

eggs. The eggs and meat are rich in proteins and are the best nutrients.

• Animals useful for agriculture and transport:- Oxen are helpful in ploughing the land. Camels and horses are also used in ploughing. The horns and the skin of animals are very useful. The dung and urine of many animals are useful in increasing the fertility of the soil.

Animal Breeding–The animal breeders generally use two methods of breeding:

• Crossing the individual animals within the same breed.

• Crossing a breed with the individual of another breed. In interbreeding, the animals chosen are those which have a good records or pedigree.

• Line Breeding:- The Line Breeding is the crossing among groups of animals with a gap of three or four generations. This yields better results than breading within the same generation.

• Out Breeding:- The animals belong to the same breed, the crossing will be between un-related groups.

• Mating outside the breed:- This is of three types:- (a) cross between different species (b) cross between individuals of different breeds. (c) Cross between a true breed and a mixed breed. This also called Grading.

• Grading:- This method is used to increase the production of meat, milk, eggs etc.

Management of Cattle:- Development pertaining to Animal Husbandry in India is hindered by the following reasons.

• Unsuitable climate and surroundings.

• Low breeding capacity.

• Insufficient nutritive feed.

Viral Diseases : (a) Cows, Buffaloes, sheep, goats and fowls are infected with small-pox. (b)Viral skin infection in sheep and goats.

Bacterial and Fungal Diseases: (a) the fowls are infected with Tuberculosis.(b) Cholera among fowls (c) Infections causing abortion (d) Diarrohoea among cattle. (e) Necrosis disease affecting the hoofs and tail. (f)Infections of feet in sheep.

Nutrition:- Animals should be fed with food rich in nutritive contents. The feed should contain carbohydrates, proteins, fats, minerals and vitamins. the feed is of two types are.

- Roughages:- It are nothing but the hay or the grassy material of paddy, wheat, maize etc. This is
- Concentrates:- It are the seeds of cotton, groundnuts, pulses, bran etc. The nutritive contents of concentrates are more than those of roughages.

Cattle Sheds:- Cattle Sheds should be constructed to protect cattle from heat, cold, and rains. Cattle well maintained yield more milk. the cattle sheds should have plenty of ventilation and drainage. Clean water should be provided to the cattle.

Effects of Pesticides

In agricultural sector and in vector borne disease control, pesticdes can cause both illnesses and death in humans. These problems arise from various circum stances, either direct or indirect human contact with pesticides.

The group of people most susceptible to direct contact with such chemicals are workers constantly exposed to or work with pesticides such as those in the agricultural or pesticide manufacturing sector. Workers who mix, load and apply pesticides are exposed to the concentrated forms of pesticides and the primary route of exposure is the skin. If these workers are not equipped with protective clothing during the application of the chemicals, absorption of pestocides through the skin would be significant.

Once in contact with the skin, pesticides can either be absorbed into the body or localised to the skin only. The common local effect seen are skin problems such as determatitis. Absorption into the body can lead to various health problems eye irritation to upper respiratory tract problems to systemic poisoning, which can lead to death.

Indirect contact with pesticides comes from the ingestion of pesticide comes from the ingestion of pesticide residues in food which would lead to an increased level of pesticide in the body is normally associated with long - term exposure to pesticides and thus may or may not be associated with illness.

The human body is a vasity complex biochemical organism, finely tuned and adaptable. It contain many different regulatory systems to make sure that things work properly in response to external conditions.

This type of regulation, known as homeostasis, occurs for all bodily processes and usually withoutany awareness or thought on our part. When external circumstances (like extreme heat or cold) or internal condition (disease or poisoning) cannot be adjusment by normal mechanisms, the signs of discomfort and disease appear.

The types of physical effects seen or felt (signs and symptoms) depend on the types of stress to which the body has been exposed. Because there are so many complex inter-relationships between the systems within the body, a single change in any system may result in numerous effects in other systems.

A body's homeostasis can be upset by physical chemical and/ or biological body's reaction to prologed stress depends on the nature of the agent as well as the degree and duration of stress. When the stress is too strong or too long and homeostass cannot be maintained or restored, diasease occurs. Poisoning by chemical agents is nothing more than chemically induced diasease and the symtoms of chemical poisoning often are the same as symtoms caused by biological agents such as bacteria or viruses.

To better understand how disease is caused by exposure to toxic chemicals, we must first understand how pesticides work within the body.

How Pesticides work

Pesticides work by changing the speed of different body funtions, increasing them (eg increasing the heart rate or sweating) or decreasing them (sometimes to the point of stopping them entirely, like breathing)

For instance, people poisoned by malathion, an insecticide, may experience increased sweating. This results in a series of event in the body as a response to the chemical. First is the biochemical inactivation of an enzyme. This (1) biochemical change leads to a (2) cellular change (in this case an increase in nerve activitay). The cellular change is then responsible for (3) physiological changes, which are the symtoms of poisoning seen or felt in particular organ systems (in this case the sweat glands). The basic progression of effects from biochemical to cellular to physiological occurs in most cases of poisoning.

Depending on the speccific biochemical mechanism of action, a poison may have very widespread effects throughout the body, or may cause a very limited change in physiological functioning in a particular region or organ. Malathion causes a very simple inactivation of an enzyme which is involve in communication between nerves. The enzyme which parathion, another type of pesticide, inactivates is, however, widespread in the body and thus a variety of effects on many body systems are seen besides sweating.

Toxicity

Toxicity is a general term used to indicate adverse effects produced by pesticides. These adverse effects camerange from slight symtoms such as headache and nausea to severe ones such as coma, convulsions and even death.

Toxicity is normal divided into four types, based on the amount of exposure to a pesticide and the time it takes for toxic symptoms to develop. The acute toxicity and chronic toxicity. The former is due to short-term exposure and happens within a relatively short period of times whereas the latter is due to long-term exposure and happens over a longer period.

Most toxic effects are reversible and do not cause permanent damage but complete recovery may take a long time. However, some pesticides may cause irreversible or permanent damage.

Pesticides can effects just one particular organ system or they may produce generalised toxicity by effecting a number of systems. Usually, the type of toxicity is subdivided intio categories based on the major organ systems affected.

Because the body only has a certain number of responses to chemical and biological stressors, it is a complicated process sorting out the signs and symtoms and determining the actual cause of human disease or illness. In many cases, it is imposible to determine whether an illness was caused by chemical exposure or by a biological agent

A history of exposure to a chemical is one important clue in helping to establish the cause of illness. However, such history does not constitute conclusive evidence that the chemical was the cause. To establish this cause effect relationship, it is important that the chemical be detected in the body (such as in the blood stream) at level known to cause illness. If the chemical produces a specific and easily detected biochemical effect (like the inhibition of the enzyme acetyl-cholinesterase), the resulting biochemical change in the body may be used as conclusive evidence. People handling chemicals frequently in the course of their jobs who become ill and need medical attention should tell their physician about their previous exposure to chemicals.

Factors determining the outcome of pesticide poisoning

If absorption of pesticide is an established fact the possibility that the absorbed dosage will be sufficient to cause death depends on the inherent toxicity of the compound.

Various ways of measuring toxicity have been developed to measure this property which include the mean **Lethal Dose** or **LD50.** The term LD50 describes acute oral dermal toxicity needed to kill at least 50% of the total test animals. Usually, information derived from the test studies would provide classification of pesticide according to the degree of toxicity with the more toxic pesticides having the smallest LD50. However, it must said that the value from LD50 should be interpreted cautiously since there is a wide difference in response between different species of organic life.

The dosage of the pesticide determines, to a large extent, the outcome of poisoning caused by them. If a large dosage has been absorbed, the poisoning that arises would be serious for not only compounds of relatively high toxicity but also to compounds of low toxicity. The control of dosage is the basis of safety in the use of pesticides. It is important to note that a sufficiently large dose of an ordinarily harmless material is fatal. On the other hand, a sufficiently small dosage of the most toxic pesticide is without effect.

The toxic manifestations of a pesticide may vary depending not only on the dose but also on the duration of the exposure. For many pesticides, the toxic effects observed from a single exposure may be quite different from that of repeated exposure. Many symtoms of repeated exposures are slow to develop and in some instances may mimic symtoms of other chronic disease, making it difficult to differentiate between a poisoning or a disease.

The final determinant on the outcome of pesticide poisoning is the route of exposure into the body. Pesticide exposure occurs through ingestion, inhalation or absorption through the skin. Certain chemicals, for instance malathion, are equally toxic by all three routes of exposure while the majority are not equally toxic by all three routes of exposure, irrespective of the duration and the dosage of exposure.

The dermal route represents the most common way for exposure and toxic effects would normally be seen if significant amounts of the pesticide have been absorbed into the body. On the whole, the number of pesticides known to be absorbed to a great extent through the skin is quite small. The efects seen in most number of cases are usually confined to the skin with symtoms such as irritation and skin sensitisation.

Use of Exececssive Chemical Fertilizers

The high solubilities of fertilizers exacerbate their tendency to degrade ecosystems, particularly through eutrophication.The use of ammonium nitrate in fertilisers is particularly damaging - plants absorb the ammonium ion preferentially to the nitrate ion - this means that the nitrate ions are not absorbed and therefore are free to be dissolved by rain leading to eutrophication.

Storage and application of some nitrogen fertilizers in some weather or soil conditions can cause emissions of the greenhouse gas nitrous oxide (N_2O). Ammonia gas (NH_3) may be emitted following application of inorganic fertilizers, or manure or slurry. Besides supplying nitrogen, ammonia can also increase soil acidity (lower pH, or "souring"). Excessive nitrogen fertilizer applications can also lead to pest problems by increasing the birth rate, longevity and overall fitness of certain pests.

The concentration of up to 100 mg/kg of cadmium in phosphate minerals (for example, minerals from Nauru and the Christmas islands) increases the contamination of soil with cadmium, for example in New Zealand. Uranium is another example of a contaminant often found in phosphate fertilizers, also radioactive Polonium-210 contained in phosphate fertilizers is absorbed by the roots of plants and stored in it tissues. Tobacco derived from plants fertilzed by rock phosphates contains Polonium-210 which emits alpha radiation estimated to cause about 11,700 lung cancer deaths each year worldwide.

For these reasons, it is recommended that knowledge of the nutrient content of the soil and nutrient requirements of the crop are carefully balanced with application of nutrients in inorganic fertilizer especially. This process is called nutrient budgeting. By careful monitoring of soil conditions, farmers can avoid wasting expensive fertilizers, and also avoid the potential costs of cleaning up any pollution created as a byproduct of their farming.

It is also possible to over-apply organic fertilizers; however, their nutrient content, their solubility, and their release rates are typically much lower than chemical fertilizers.[*citation needed*] By their nature, most organic fertilizers also provide increased physical and biological storage mechanisms to soils, which tend to mitigate their risks.

The use of fertilizers on a global scale emits significant quantities of greenhouse gas into the atmosphere. Emissions come about through the use of:

- animal
- manures and
- urea, which release
- methane,
- nitrous oxide,

- ammonia, and
- carbon dioxide in varying quantities depending on their form (solid or liquid) and management (collection, storage, spreading)
- fertilizers that use
- nitric acid or
- ammonium bicarbonate, the production and application of which results in emissions of
- nitrogen oxides,
- nitrous oxide,
- ammonia and
- carbon dioxide into the atmosphere.

By changing processes and procedures, it is possible to mitigate some, but not all, of these effects on anthropogenic climate change.

The nitrogen-rich compounds found in fertilizer run-off is the primary cause of a serious depletion of oxygen in many parts of the ocean, especially in coastal zones; the resulting lack of dissolved oxygen is greatly reducing the ability of these areas to sustain oceanic fauna.

9

BIODIVERSITY

Biodiversity is the variation of life forms within a given ecosystem, biome, or for the entire Earth. Biodiversity is often used as a measure of the health of biological systems. The biodiversity found on Earth today consists of many millions of distinct biological species, which is the product of nearly 3.5 billion years of evolution

Definitions

The most straightforward definition is "variation of life at all levels of biological organization". A second definition holds that biodiversity is a measure of the relative diversity among organisms present in different ecosystems. "Diversity" in this definition includes diversity within a species and among species, and comparative diversity among ecosystems.

A third definition that is often used by ecologists is the "totality of genes, species, and ecosystems of a region". An advantage of this definition is that it seems to describe most circumstances and present a unified view of the traditional three levels at which biodiversity has been identified:

- genetic diversity

- species diversity
- ecosystem diversity

Genetic diversity is a level of biodiversity that refers to the total number of genetic characteristics in the genetic makeup of a species. It is distinguished from genetic variability, which describes the tendency of genetic characteristics to vary.

The academic field of population genetics includes several hypothess regarding genetic diversity. The neutral theory of evolution proposes that diversity is the result of the accumulation of neutral substitutions

Species diversity refers to the number and distribution of species in one location. Simply the measure of the number of different species within a given area. Humans have a huge effect on species diversity; the main reasons are:

- Destruction, Modification, and/or Fragmentation of Habitat
- Introduction of Exotic Species
- Overharvest
- Global Climate Change

Ecosystem diversity refers to the diversity of a place at the level of ecosystems. It is contrasted with biodiversity, which refers to variation in species rather than ecosystems.

Biodiversity is the variety and differences among living organisms from all sources, including terrestrial, marine, and other aquatic ecosystems and the ecological complexes of which they are a part. This includes genetic diversity within and between species and of ecosystems. Thus, in essence, biodiversity represents all life.

Diversity of Ecosystem

An ecosystem is a community plus the physical environment that it occupies at a given time. An ecosystem can exist at any scale, for example, from the size of a small tide pool up to the size of the entire biosphere. However, lakes, marshes, and forest stands represent more typical examples of the areas that are compared in discussions of ecosystem diversity.

Broadly speaking, the diversity of an ecosystem is dependent on the physical characteristics of the environment, the diversity of species present, and the interactions that the species have with each other and with the environment. Therefore, the functional complexity of an ecosystem can be expected to increase with the number and taxonomic diversity of the species present, and the vertical and horizontal complexity of the physical environment. However, one should note that some ecosystems (such as submarine black smokers, or hot springs) that do not appear to be physically complex, and that are not especially rich in species, may be considered to be functionally complex. This is because they include species that have remarkable biochemical specializations for surviving in the harsh environment and obtaining their energy from inorganic chemical sources

The physical characteristics of an environment that affect ecosystem diversity are themselves quite complex. These characteristics include, for example, the temperature, precipitation, and topography of the ecosystem. Therefore, there is a general trend for warm tropical ecosystems to be richer in species than cold temperate ecosystems. Also, the energy flux in the environment can significantly affect the ecosystem. An exposed coastline with high wave energy will have a considerably different type of ecosystem than a low-energy environment such as a sheltered salt marsh.

Types of Species

Alpha diversity : Alpha diversity (α-diversity) is the biodiversity within a particular area, community or ecosystem, and is usually expressed as the Species richness of the area. This can be measured by counting the number of taxa (distinct groups of organisms) within the ecosystem (eg. families, genera, species). However, such estimates of species richness are strongly influenced by sample size, so a number of statistical techniques can be used to correct for sample size to get comparable values.

Beta Diversity : Beta diversity (β-diversity) is a measure of biodiversity which works by comparing the species diversity between ecosystems or along environmental gradients. This involves comparing the number of taxa that are unique to each of the ecosystems.It is the rate of change in species composition across habitats or among communities. It gives a quantitative measure of diversity of communities that experience changing environments.For example, the beta diversity between the woodland and the hedgerow habitats is 7 (representing the 5 species found in the woodland but not the hedgerow, plus the 2 species found in the hedgerow but not the woodland). Thus, beta diversity allows us to compare diversity between ecosystems.

Gamma diversity (â-diversity) is a measure of biodiversity. It refers to the total biodiversity over a large area or region. It is the total of á and â diversity.

Hypothetical species Open field habitat	Woodland habitat	Hedgerow habitat
A	X	
B	X	
C	X	
D	X	
E	X	
F	X	X
G	X	X
H	X	X
I	X	X
J	X	X
K	X	
L	X	X
M		X
N		X
Alpha diversity	**10**	
Beta diversity	**73**	
Woodland vs. hedgerow:	**7**	
Hedgerow vs. open field:	**8**	
Woodland vs. open field:	**13**	
Gamma diversity	**14**	

According to Whittaker (1972), gamma diversity is the richness in species of a range of habitats in a geographic area (e.g.,a landscape, an island) and it is consequent on the alpha diversity of the individual communities and the range of differentiation or beta diversity among them. Like alpha diversity, it is a quality which simply has magnitude, not direction and can be represented by a single number (a scalar).

The internal relationship between alpha, beta and gamma diversity can be represented as

$$\beta = \gamma / \alpha$$

Gamma diversity is a measure of the overall diversity for the different ecosystems within a region.

Defines gamma diversity as "geographic-scale species diversity". In the example in Table 1, the total number of species for the three ecosystems 14, which represent the gamma diversity.

Alpha, beta and gamma diversity for hypothetical species of birds in three different ecosystems

Genetic Diversity : Genetic diversity is a level of biodiversity that refers to the total number of genetic characteristics in the genetic makeup of a species. It is distinguished from genetic variability, which describes the tendency of genetic characteristics to vary.

The academic field of population genetics includes several hypotheses regarding genetic diversity. The neutral theory of evolution proposes that diversity is the result of the accumulation of neutral substitutions. Diversifying selection is the hypothesis that two subpopulations of a species live in different environments that select for different alleles at a particular locus. This may occur, for instance, if a species has a large range relative to the mobility of individuals within it.

Genetic diversity refers to any variation in the nucleotides, genes, chromosomes, or whole genomes of organisms (the genome is the entire complement of DNA within the cells or organelles of the organism). Genetic diversity at its most elementary level is represented by differences in the sequences of nucleotides (adenine, cytosine, guanine, and thymine) that form the DNA (deoxyribonucleic acid) within the cells of the organism. The DNA is contained in the chromosomes present within the cell; some chromosomes are contained within specific organelles in the cell (for example, the chromosomes of mitcchondria and chloroplast).

Nucleotide variation is measured for discrete sections of the chromosomes, called genes. Thus, each gene compromises a hereditary section of DNA that occupies a specific place of the chromosome, and controls a particular characteristic of an organism.

Most organisms are diploid, having two sets of chromosomes, and therefore two copies (called alleles) of each gene. However, some organisms can be haploid, triploid, or tetraploid (having one, three, or four sets of chromosomes respectively). Within any single organism, there may be variation between the two (or more) alleles for each gene. This variation is introduced either through mutation of one of the alleles, or as a result of sexual reproduction. During sexual reproduction, offspring inherit alleles from both parents and these alleles might be slightly different, especially if there has been migration or hybridization of organisms, so that the parents may come from different populations and gene pools. Also, when the offspring's chromosomes are copied after fertilization, genes can be exchanged in a process called sexual recombination.

Each allele codes for the production of amino acids that string together to form proteins. Thus differences in the nucleotide sequences of alleles result in the production of slightly different strings of amino acids or variant forms of the proteins. These proteins code for the development of the anatomical and physiological characteristics of the organism, which are also responsible for determining aspects of the behaviour of the organism.

Genetic diversity is, therefore, a key component for conservation efforts associated with population management. The genetic constitution of an organism the arrangement of the DNA into genes on the chromosomes is also referred to as its genotype. Hence, variation that exists within the genetic constitution of an organism is often referred to as genotypic variation .

Floristic Regions of India

The forests of the Himalayan region (from Kashmir to Kumaon) possess a rich variety of Deodar, Blue Pine, Spruce, Silver Fir, Chir, Pine, other Conifers and broad leaved temperate trees. The Eastern Himalayas (from Sikkim to Darjeeling) forests has Oaks, Laurela, Maples, Rhododendroms, Alder, Birth, Juniper

Trees. The Assam region is famous for its evergreen forests. Malabar region, besides the rich forest, produces important commercial crops such as coconut, betel nut, pepper, coffee, tea, rubber, cardamom etc. The Andaman region abounds in evergreen, mangrove, beach and diluvial forests.

There are 14 major vegetation types identified in India. They are Tropical wet evergreen, Tropical semi evergreen, Tropical moist deciduous, Tropical dry deciduous, Tropical thorn, Tropical desert, very dry ever green, Subtropical wet, Subtropical pine, Subtropical dry, wet temperate, moist temperate, The Alpine, The Tidal Forest.

India can be divided into eight distinct-floristic-regions, namely, the western Himalayas, the eastern Himalayas, Assam, the Indus plain, the Ganga plain, the Deccan, Malabar and the Andamans.

The Western Himalayan region extends from Kashmir to Kumaon. Its temperate zone is rich in forests of *chir*, pine, other conifers and broad-leaved temperate trees. Higher up, forests of *deodar,* blue pine, spruce and silver fir occur. The alpine zone extends from the upper limit of the temperate zone of about 4,750 metres or even higher. The characteristic trees of this zone are high-level silver fir, silver birch and junipers. The eastern Himalayan region extends from Sikkim eastwards and embraces Darjeeling, Kurseong and the adjacent tract. The temperate zone has forests of oaks, laurels, maples, rhododendrons, alder and birch. Many conifers, junipers and dwarf willows also occur here. The Assam region comprises the Brahmaputra and the Surma valleys with evergreen forests, occasional thick clumps of bamboos and tall grasses. The Indus plain region comprises the plains of Punjab, western Rajasthan and northern Gujarat. It is dry and hot and supports natural vegetation. The Ganga plain region covers the area which is alluvial plain and is under cultivation for wheat, sugarcane and rice. Only small areas support forests of widely differing types. The Deccan region comprises the entire tableland of the Indian Peninsula and supports vegetation of various kinds from scrub jungles to mixed deciduous forests. The Malabar region covers the excessively humid belt of mountain country parallel to the west coast of the Peninsula. Besides being rich in forest vegetation, this region produces important commercial crops, such as coconut, betel nut, pepper, coffee and tea, rubber and cashew nut. The Andaman region abounds in evergreen, mangrove, beach and diluvia forests. The Himalayan region extending from Kashmir to Arunachal Pradesh through Nepal, Sikkim, Bhutan, Meghalaya and Nagaland and the Deccan Peninsula is rich in endemic flora, with a large number of plants which are not found elsewhere.

The value of Biodiversity

Biological resources provide the basis for life on earth, including that of humans. The fundamental social, ethical, cultural, and economic values of these resources have been recognized in religion, art, and literature from the earliest days of recorded history. The great interest that children have in nature, the numerous wildlife clubs, the generous donations made to non-governmental conservation organizations, the political support for "Green Parties," and the popularity of zoos and wildlife films are economic expressions of preference and show that the general public does not think of biological resources merely in terms of cash value.

But in order to compete for the attention of government and commercial decision-makers in today's world, policies regarding biological diversity first need to demonstrate in economic terms the contribution biological resources make to the country's social and economic development. Even partial valuation in monetary terms of the benefits of conserving biological resources can provide at least a lower limit to the full range of benefits and demonstrate the conservation can yield a profit in terms that are meaningful to national accounts.

Three main approaches have been used for determining the value of biological resources:

- *assessing the value of nature's products* — such as firewood, fodder, and game meat—that are consumed directly, without passing through a market ("consumptive use value");
- *assessing the value of products* that are commercially harvested, such as timber, fish, game meat sold in a market, ivory, and medicinal plants ("productive use value"); and
- *assessing indirect values* of ecosystem functions, such as watershed protection, photosynthesis, regulation of climate, and

production of soil ("non-consumptive use value"), along with the intangible values of keeping options open for the future ("option value") and simply knowing that certain species exist ("existence value").

Threats to Biodiversity

Extinction is a natural event and, from a geological perspective, routine. We now know that most species that have ever lived have gone extinct. The average rate over the past 200 my is 1-2 species per year, and 3-4 families per my. The average duration of a species is 2-10 million years (based on last 200 million years). There have also been occasional episodes of mass extinction, when many taxa representing a wide array of life forms have gone extinct in the same blink of geological time.

In the modern era, due to human actions, species and ecosystems are threatened with destruction to an extent rarely seen in earth history. Probably only during the handful of mass extinction events have so many species been threatened, in so short a time.

What are these human actions? There are many ways to conceive of these - let's consider two. First, we can attribute the loss of species and ecosystems to the accelerating transformation of the earth by a growing human population. As the human population passes the six billion mark, we have transformed, degraded or destroyed roughly half of the word's forests. We appropriate roughly half of the world's net primary productivity for human use. We appropriate most available fresh water, and we harvest virtually all of the available productivity of the oceans. It is little wonder that species are disappearing and ecosystems are being destroyed. Second, we can examine six specific types of human actions that threaten species and ecosystems - the "sinister sextet"

Over-hunting has been a significant cause of the extinction of hundreds of species and the endangerment of many more, such as whales and many African large mammals. Most extinctions over past several hundred years are mainly due to over-harvesting for food, fashion, and profit.

Commercial hunting, both legal and illegal (poaching), is the principal threat. Snowy egret, passenger pigeon, heath hen are USA examples. At $16,000 per pound, and $40,000 to $100,000 per horn, it is little wonder that some rhino species are down to only a few thousand individuals, with only a slim hope of survival in the wild. The pet and decorative plant trade falls within this commercial hunting category, and includes a mix of legal and illegal activities. The annual trade is estimated to be at least $5 billion, with perhaps 1/4 to 1/3 of it illegal.

Sport or recreational hunting causes no endangerment of species where it is well regulated, and may help to bring back a species from the edge of extinction. Many wildlife managers view sport hunting as the principal basis for protection of wildlife.

While over-hunting, particularly illegal poaching, remains a serious threat to certain species, for the future, it is less important than other factors mentioned next.

Habitat loss/degradation/fragmentation is an important cause of known extinctions. As deforestation proceeds in tropical forests, this promises to become the cause of mass extinctions caused by human activity.

All species have specific food and habitat needs. The more specific these needs and localized the habitat, the greater the vulnerability of species to loss of habitat to agricultural land, livestock, roads and cities. In the future, the only species that survive are likely to be those whose habitats are highly protected, or whose habitat corresponds to the degraded state associated with human activity.

Habitat damage, especially the conversion of forested land to agriculture (and, often, subsequent abandonment as marginal land), has a long human history. It began in China about 4,000 years ago, was largely completed in Europe by about 400 years ago, and swept across USA over the past 200 years or so. Viewed in this historical context, we are now mopping up the last forests of Pacific Northwest.

In the new world tropics, lowland, seasonal, deciduous forest began to disappear after 1500 with Spanish and Portuguese colonization of the New World. These were the forested regions most easily converted to agriculture, and with a more welcoming climate. The more forbidding, tropical humid forests came under attack mainly in 20th Century, under the combined influences of population growth, inequitable land and income distribution, and development policies that targeted rain forests as the new frontier to colonize.

Tropical forests are so important because they harbor at least 50%, and perhaps more, of world's biodiversity. Direct observations, reinforced by satellite data, documents that these forests are declining. The original extent of tropical rain forests was 15 million km^2. Now there remains about 7.5-8 million km^2, so half is gone. The current rate of loss is estimated at near 2% annually (100,000 km^2 destroyed, another 100,000 km^2 degraded). While there is uncertainty regarding the rate of loss, and what it will be in future, the likelihood is that tropical forests will be reduced to 10-25% of their original extent by late 21st C. Habitat fragmentation is a further aspect of habitat loss that often goes unrecognized. The forest, meadow, or other habitat that remains generally is in small, isolated bits rather than in large, intact units. Each is a tiny island that can at best maintain a very small population. Environmental fluctuations, disease, and other chance factors make such small isolates highly vulnerable to extinction. Any species that requires a large home range, such as a grizzly bear, will not survive if the area is too small. Finally, we know that small land units are strongly affected by their surroundings, in terms of climate, dispersing species, etc. As a consequence, the ecology of a small isolate may differ from that of a similar ecosystem on a larger scale.

For the future, habitat loss, degradation, and fragmentation combined is the single most important factor in the projected extinction crisis.

India as a Mega Diversity Nation

India is one of the 12-mega biodiversity countries of the world. With only 2.4% of the land area, India already accounts for 7-8% of the recorded species of the world. Over 46,000 species of plants and 81,000 species of animals have been recorded in the country so far by the Botanical Survey of India, and the Zoological Survey of India, respectively. India is an acknowledged centre of crop diversity, and harbours many wild relatives and breeds of domesticated animals.

Name the mega biodiversity countries : Biodiversity is not equally distributed all over the globe. Certain countries are characterized by high species richness and more number of endemic species. These countries are known as Mega biodiversity countries. Twelve such countries have been identified. Together, these countries harbour 60- 70% of the world's recorded biodiversity. These countries are: Brazil, Colombia, Ecuador, Peru, Mexico, Madagascar, Zaire, Australia, China, India, Indonesia and Malaysia.

Hot-spots of Biodiversity in World

A **biodiversity hotspot** is a biogeographic region with a significant reservoir of biodiversity that is threatened with destruction.

The biodiversity hotspots by region

North and Central America

- California floristic province
- Caribbean Islands
- Madrean pine-oak woodlands
- Mesoamerica

South America

- Atlantic Forest
- Cerrado
- Chilean Winter Rainfall-Valdivian Forests
- Tumbes-Chocó-Magdalena
- Tropical Andes

Europe and Central Asia

- Caucasus
- Irano-Anatolian
- Mediterranean Basin
- Mountains of Central Asia

Africa

- Cape Floristic Region
- Coastal forests of eastern Africa
- Eastern Afromontane
- Guinean Forests of West Africa
- Horn of Africa
- Coastal Forests of Eastern Africa

- Madagascar and the Indian Ocean Islands
- Maputaland-Pondoland-Albany
- Succulent Karoo

Asia-Pacific

- East Melanesian Islands
- Eastern Himalaya
- Indo-Burma
- Japan
- Mountains of Southwest China
- New Caledonia
- New Zealand
- Philippines
- Polynesia-Micronesia
- Southwest Australia
- Sundaland
- Wallacea
- Western Ghats and Sri Lanka

Critiques of Hotspots

The high profile of the biodiversity hotspots approach has resulted in considerable criticism. Papers such as Kareiva & Marvier (2003) have argued that the biodiversity hotspots:

- Do not adequately represent other forms of species richness (e.g. total species richness or threatened species richness).
- Do not adequately represent taxa other than vascular plants (e.g. vertebrates, or fungi).
- Do not protect smaller scale richness hotspots.
- Do not make allowances for changing land use patterns. Hotspots represent regions that have experienced considerable habitat loss, but this does not mean they are experiencing on going habitat loss. On the other hand, regions that are relatively intact (e.g. the Amazon Basin) have experienced relatively little land loss, but are currently losing habitat at tremendous rates.
- Do not protect ecosystem services
- Do not consider phylogenetic diversity.

A recent series of papers has pointed out that biodiversity hotspots (and many other priority region sets) do not address the concept of cost. The purpose of biodiversity hotspots is not simply to identify regions that are of high biodiversity value, but to prioritize conservation spending. The regions identified include regions in the developed world (e.g. the California Floristic Province), alongside regions in the developing world (e.g. Madagascar). The cost of land is likely to vary between these regions by an order of magnitude or more, but the biodiversity hotspots do not consider the conservation importance of this difference.

Endangered and Endemic Species of India

Endangered Species : According to official records, there are more than 130000 animal species in India. According to some, the number may actually be much more than this. There has been a threat to this natural treasure with the increasing destruction of their habitat like the tropical forests and biosphere reserves. There are a number of causes which lead to the endangerment of a particular species such as habitat destruction, paucity of prey, overexploitation etc. The animal population has been declining by every passing minute and we are likely to face grave consequences until we do not react fast. Moreover, the plants and animals hold immense medicinal, agricultural, ecological and commercial value. This is the time when the endangered species must be protected so that our future generations are not deprived of experiencing this valuable treasure. Some of the animal species classified as endangered are:

Great Indian One-horned Rhinoceros : The Indian Rhinoceros has been hunted for its horn, which is believed to have aphrodisiac properties and can be as long as 53 cm. There are only about 1500 of these animals left in the wild which are mainly found in the Kaziranga and Orang national parks.

Asian White Backed Vulture : These creatures, seen all over the subcontinent in large numbers until the ninety's are on the verge of extinction because of reasons such as poisoning due to feeding on dead cattle treated with certain anti-inflammatory drugs.

Asian Elephant : Loss of habitat and poaching has led to the near extinction of this gentle giant and there are only about 15 to 20 thousand of these creatures left in the wild. The Asian Elephant is mainly found in the terai region of Uttar Pradesh, Bengal, Assam, Kerala and Karnataka.

Leopard : There has been much said about saving the Tiger, but far less attention has been paid to its dotted counterpart, which is equally threatened and there are only about 7500 leopards to be seen around in the Indian jungles. These are concentrated in parts of central India and north east.

Asiatic Lion : Found in the Gir national park, on last count there are only about 352 of these ferocious beasts left. Now is the time we should act and protect the Lion if we want our children to see and admire this marvellous creature.

Snow Leopard : The exotic looking creature is mostly seen in the upper reaches of the Himalayas. Poaching and paucity of prey have led to its near extinction, but thanks to environmentalist who woke up to this threat and took immediate measures for the animal's survival.

Endemic species : Endemism is the ecological state of being unique to a place. Endemic species are not naturally found elsewhere. The place must be a discrete geographical unit, such as an island, habitat type, nation, or other defined area or zone. For example, the Orange-breasted Sunbird is endemic to Fynbos, meaning it is exclusively found in the Fynbos vegetation type of southwestern South Africa.

There are two subcategories of endemism - *paleoendemism and neoendemism*. Paleoendemism refers to a species that was formerly widespread but is now restricted to a smaller area. Neoendemism refers to a species that has recently arisen such as a species that has diverged and become reproductively isolated, or one that has formed following hybridization and is now classified as a separate species. This is a common process in plants especially those which exhibit polyploidy.

Endemic types or species are especially likely to develop on islands because of their geographical isolation. This includes remote island groups, such as Hawaii, the Galápagos Islands, and Socotra. Endemism can also occur in biologically isolated areas such as the highlands of Ethiopia, or large bodies of water like Lake Baikal.

Endemics can easily become endangered or extinct because of their restricted habitat and vulnerability to the actions of man, including the introduction of new organisms. There were millions of both Bermuda Petrels and "Bermuda cedars" (actually *junipers*) in Bermuda when it was settled at the start of the 17th century. By the end of the century, the petrels were thought to be extinct. Cedars, decimated by centuries of shipbuilding, were driven nearly to extinction in the 20th Century by the introduction of a parasite. Both petrels and cedars are very rare today, as are other species endemic or native to Bermuda.

Endemic organisms are not the same as indigenous organisms — a species that is indigenous to somewhere may be native to other locations as well. An introduced species, also known as a naturalized or exotic species, is an organism that is not indigenous to a given place or area.

Biodiversity and Its Conservation : Biotechnology involves the use of all life forms for human welfare. Therefore, extinction of wild species and destruction of ecosystems has been a major concern of policy makers and biotechnologists alike. One of the major efforts has been to conduct a survey and conserve country's biodiversity, so as to save wild plants and animals from extinction.

National parks and sanctuaries have been established in many countries to meet this objective. Under the auspices of the United Nations also, funds are being established and other efforts being made for conservation of germplasm at the global level. Biodiversity studies thus include the following:

(i) a systematic examination of the full array of organisms on this globe and

(ii) a study of the methods by which diversity can be maintained and used for the benefit of mankind.

Ex Situ Conservation -Ex situ conservation, using sample populations is done through establishment of 'gene banks', which include genetic resource centres, zoos, botanical gardens, culture collections, etc. Although the phrase 'gene banks' often refers to only ex situ conservation facilities, they do include in-situ conservation methods, which include national parks and sanctuaries.

Ex-situ conservation means literally, "off-site conservation". It is the process of protecting an endangered species of plant or animal by removing part of the population from a threatened habitat and placing it in a new location, which may be a wild area or within the care of humans. While ex-situ conservation comprises some of the oldest and best known conservation methods, it also involves newer, sometimes controversial laboratory methods.

This has become particularly important for conservation of crop varieties and wild genetic resources, because of their utility in future crop improvement and afforestation programmes. However, there has been a competition for relative allocation of efforts directed towards in situ and ex situ conservation.

For instance in 1987 for ex situ programmes, USA allowed only 1% of a total of 37.5 million dollars meant for biodiversity conservation, excluding contributions to international systems of gene banks.

In Situ Conservation - This type of conservation applies only to wild fauna and flora and not to the domesticated animals and plants, because conservation is achieved by protection of populations in nature. It includes a system of protected areas of different categories, e.g. National Parks, Sanctuaries, Nature Reserves, Natural Monuments, Cultural Landscapes, Biosphere Reserves, etc.

In-situ conservation means "on-site conservation". It is the process of protecting an endangered plant or animal species in its natural habitat, either by protecting or cleaning up the habitat itself, or by defending the species from predators. Increasingly, this term is also being applied to the conservation of agricultural biodiversity in agro ecosystems by farmers, especially those using unconventional farming practices. One benefit to *in-situ* conservation is that it maintains recovering populations in the surrounding where they have developed their distinctive properties. Another is that this strategy helps ensure the ongoing processes of evolution and adaptation within their environments. As a last resort, ex-situ conservation may be used on some or all of the population, when *in-situ* conservation is too difficult, or impossible.

Wildlife and livestock conservation is mostly based on *in situ* conservation. This involves the protection of wildlife habitats. Also, sufficiently large reserves are maintained to enable the target species to exist in large numbers. The population size must be sufficient to enable the necessary genetic diversity to survive within the population, so that it has a good chance of continuing to adapt and evolve over time. This reserve size can be calculated for target species by examining the population density in naturally-occurring situations. The reserves must then be protected from intrusion or destruction by man, and against other catastrophes.

In situ conservation of forest trees. Efforts are being planned by CGIAR's latest international agricultural research centre, CIFOR (Centre for International Forestry Research) to conduct research for in situ conservation of forest trees. It is argued that although the tropical foresters reserve forest areas against other land uses, these have been used for multiple purposes (for water catchment, protection of mammals or birds, etc.).Therefore, CIFOR will conduct research and find more rational location of in situ reserves for the conservation of woody plants germplasm. This work will be done in close collaboration with IUCN (International Union for Conservation of Nature and Natural Resources), and the national institutes in developing countries.

National Parks

A **national park** is a reserve of land, usually declared and owned by a national government, protected from most human development and pollution. National parks are protected areas of IUCN category II. The largest national park in the world is the Northeast Greenland National Park, which was established in 1974.

National parks are usually located in places which have been largely undeveloped, and often feature areas with exceptional native animals, plants and ecosystems (particularly endangered examples of such), biodiversity, or unusual geological features. Occasionally, national parks are declared in developed areas with the goal of returning the area to resemble its original state as closely . s possible. In some countries, such as England and Wales, areas designated as a national park are not wilderness, nor owned by the government, and can include substantial settlements and land uses which are often integral parts of the landscape.

Wildlife Santuaries

India is unique in the richness and diversity of its vegetation and wildlife. India's national parks and **wild life sanctuaries** (including bird sanctuaries) from Ladakh in Himalayas to Southern tip of Tamil Nadu, are outstanding and the country continues to "WOW" the tourists with its rich bio-diversity and heritage. Wildlife sanctuaries in India attracts people from all over the world as the rarest of rare species are found here. With 96 national parks and over 500 wildlife sanctuaries, the range and diversity of India's wildlife heritage is matchless. Some of the important sanctuaries in India are The Jim Corbett Tiger Reserve – Uttaranchal, Kanha National Park and Bandhavgarh National Park in Madhya Pradesh, Ranthambhor National Park - Sawai Madhopur, Gir National Park - Sasangir (Gujarat) etc. Supporting a great variety of mammals and over 585 species of birds, India's first national park, the Corbett was established in the foothills of Himalayas.

Wildlife lovers will be excited to see magnificent Bird Sanctuary at Bharatpur, Rajasthan as it is the second habitat in the world that is visited by the Siberian Cranes in winter and it provides a vast breeding area for the native water birds.

In the Indian deserts, the most discussed bird is the Great Indian bustard. In western Himalayas, one can see birds like Himalayan monal pheasant, western tragopan, koklass, white crested khalij pheasant, griffon vultures, lammergiers, choughs, ravens. In the Andaman and Nicobar region, about 250 species and sub species of birds are found, such as rare Narcondum horn bill, Nicobar pigeon and megapode.

While the national park and sanctuaries of northern and central India are better known, there are quite a few parks and sanctuaries in South India, too. For e.g., Madumalai in Tamil Nadu and Bandipur Tiger Reserve and Nagahole National Park in Karnataka.

A tour of **Indian wildlife sanctuaries** and national parks is a fabulous experience. Contrary to the African Safari, the vegetation and terrain in India is such that wild animals are often solitary or in small herds, elusive and shy. Ranges of Safari Packages are on offer, courtesy the tourism departments of states as well as tour and travel agencies. These Safari / Safari Packages are unique and unparalleled. These Safaris facilitate seeing a tiger, a rhinoceros or a herd of wild elephants.

India has unmatched variety of flora and fauna that makes it extensively different from the rest of the world. Tourists visiting for wildlife tour in India, will enjoy during any season, but to experience migrating birds, tiger, leopard, barasingha and other rare species, then winter is the best season to visit sanctuaries especially for those tourists coming for wildlife tour in India. Due to water scarcity in the hot weather, animals come out in herd in search of water; therefore most of the sanctuaries are closed during summer season. Tourists can opt for jungle safari in an open jeep but the experience on elephants back is overwhelming.

Wildlife Conservation Society(WCS) India in association with other NGO partners and tribal people, is making every possible effort to develop new models of wildlife conservation to preserve India's most treasured fauna and to protect the environment.

A **biosphere reserve** is an international conservation designation given by UNESCO under its **Programme on Man and the Biosphere (MAB)**. The World Network of Biosphere Reserves is the collection of all 531 biosphere reserves in 105 countries (as of May, 2008).

Biosphere reserves are created "to promote and demonstrate a balanced relationship between humans and the biosphere." Biosphere Reserves must "encompass a mosaic of ecological systems," and thus consist of combinations of terrestrial, coastal, or marine ecosystems.

Through appropriate zoning and management, the conservation of these ecosystems and their biodiversity is sought to be maintained.

The design of the reserve must include a legally protected core area, a buffer area where non-conservation activities are prohibited, and a transition zone where approved practices are permitted. This is done with regard for the sustainable use of natural resources for the benefit of local communities. This effort requires relevant research, monitoring, education and training.

10

ENVIRONMENTAL POLLUTION

Any undesirable change in the physical, chemical, or biolgical characteristics of the air, water, or land that can harmfully affect health, survival, or activities of humans or other living organisms, is termed as **Pollution.**

Pollution can take the form of chemical substances, or energy, such as noise, heat, or light energy. Pollutants, the elements of pollution, can be foreign substances or energies, or naturally occurring; when naturally occurring, they are considered contaminants when they exceed natural levels. Pollution is often classed as point source or non-point source pollution.

Any substitute which causes lowering the quality of environment or pollution in called **Pollutants.**

A Pollutant is defined as any form of energy or matter that causes degradation and pollution in the existing natural balance of ecosystems.

Pollutants are divided into two major types :

(1) Degradable Pollutants

(2) Non-Degradable Pollutants

Degradable Pollutants

A degradable pollutant can be decomposed, removed, or consumed and thus reduced to acceptable levels either by natural processes or by human engineered system (such as sewage treatment plants), as long as the systems are not overloaded.

There are two classes of degradable pollutants : rapidly degradable (nonpersistent) and slowly degradable (persistent).

Rapidly degradable pollutants : Rapidly degradable pollutants, such as human sewage and animal and crop wastes, and normally be decomposed rather quickly if the system is not overladed.

Slowly degradable pollutants : Slowly degradable pollutants, such as DDT and some radioactive materials, decompose slowly but eventually are either broken down completely or reduced to harmless levels. Some radioactive materials the give off harmful radiation, such as iodine-131, decay to harmless levels in a few minutes or hours and are rapidly degradable pollutants.

Non-degradable pollutants : Non-degradable pollutants are not broken down by natural processes. Examples or non-degradable pollutants are mercury, lead, and some of their compounds and some of their compounds and some plastics. Like slowly degradable pollutants, non-degradable pollutants must be either prevented from entering the air, water, and soil or kept below harmful levels by removal from the environment.

Environmental Pollution

Pollution can take many forms. The air we breathe, the water we drink, the ground we grow our food, and even the increasing noise we hear every day all contribute to health problems and a lower quality of life. Environmental pollution problems have become a global issue.

Major Categories of Environmental Pollution

These are Major categories of Environmental Pollutions are following

(1) Air Pollution

(2) Water Pollution
(3) Soil Pollution
(4) Marine Pollution
(5) Noise Pollution
(6) Theism Pollution
(7) Nuclear Pollution
(8) Solid waste Pollution

Air Pollution

Air pollution, the release of chemicals and particulates into the atmosphere. Common air pollutants include carbon monoxide, sulfur dioxide, chlorofluorocarbons (CFC) and nitrogen oxides produced by industry and motor vehicles. Photochemical ozone and smog are created as nitrogen oxides and hydrocarbons react to sunlight.

Air is very important for all types of life in the biosphere. Human life is not possible without air because man can live for a few days without water or for a few weeks without food but cannot survive even for a few minutes without air. It constitutes about 80 per cent of the total intake of all things by a person very day as a person breathes 22,000 times a day inhaling 35 gallons or 16 kilogrames of air which he obtains from the oxygen-rich atmosphere surrounding the earth.

The atmosphere is a gaseous envelope which surrounds the earth from all sides and the air is a mechanical mixture of several gases, mainly nitrogen (78.09%), oxygen (20.95%), argan (0.93%) and carbon dioxide (0.03%). Besides, other trace gases like neon, krypton, helium, hydrogen, xenon and ozone are also present.

Air pollution can cause death, impair health, reduce visibility, bring about vast econmic losses and contribute to the general deterioration of both our cities and country-side. It can also cause intangible losses to historical monuments such as the Taj Mahal which is believed to be badly affected by air pollution.

The *AIR POLLUTION* is generally accomplished through the pollutants of gases and solid and liquid particles of both organic and inorganic chemical. It is true that air is never pure because some gases such as sulphur dioxide, carbon monoxide; hydrogen suphide emissions from volcanoes, swamps, dusts, salt spray, pollens from plants, etc., are continuously added to the air by these natural processes. Thus the air becomes polluted when its natural composition is disturbed either by natural or mon-made sources or activities or by both.

Air pollution, the release of chemicals and particulates into the atmosphere. Common air pollutants include carbon monoxide, sulfur dioxide, chlorofluorocarbons (CFC) and nitrogen oxides produced by industry and motor vehicles. Photochemical ozone and smog are created as nitrogen oxides and hydrocarbons react to sunlight.

Air pollution results from a variety of causes, not all of which are within human control. Dust storms in desert areas and smoke from forest fires and grass fires contribute to chemical and particulate pollution of the air. The source of pollution may be in one country but the impact of pollution may be felt elsewhere.

Sources and Classification of Air Pollutants

Classification of Air Pollutants:

Carbon monoxide (CO) is a colourless, odourless gas that is produced by the incomplete burning of carbon based fuels including petrol, diesel, and wood. It is also produced from the combustion of natural and synthetic products such as cigarettes. It lowers the amount of oxygen that enters our blood. It can slow our reflexes and make us confused and sleepy.

Carbon dioxide (CO2) is the principle greenhouse gas emitted as a result of human activities such as the burning of coal, oil, and natural gases.

Chlorofluorocarbons (CFC) are gases that are released mainly from air-conditioning systems and refrigeration. When released into the air, CFCs rise to the stratosphere, where they come in contact with few other gases, which leads to a reduction of the ozone layer that protects the earth from the harmful ultraviolet rays of the sun.

Lead is present in petrol, diesel, lead batteries, paints, hair dye products, etc. Lead affects children in particular. It can cause nervous system damage and digestive problems and, in some cases, cause cancer.

Ozone occurs naturally in the upper layers of the atmosphere. This important gas shields the earth from the harmful ultraviolet rays of the sun. However, at the ground level, it is a pollutant with highly toxic effects. Vehicles and industries are the major source of ground-level ozone emissions. Ozone makes our eyes itch, burn, and water. It lowers our resistance to colds and pneumonia.

Nitrogen oxide (Nox) causes smog and acid rain. It is produced from burning fuels including petrol, diesel, and coal. Nitrogen oxides can make children susceptible to respiratory diseases in winters.

Suspended particulate matter (SPM) consists of solids in the air in the form of smoke, dust, and vapour that can remain suspended for extended periods and is also the main source of haze which reduces visibility. The finer of these particles, when breathed in can lodge in our lungs and cause lung damage and respiratory problems.

Sulphur dioxide (SO2) is a gas produced from burning coal, mainly in thermal power plants. Some industrial processes, such as production of paper and smelting of metals, produce sulphur dioxide. It is a major contributor to smog and acid rain. Sulfur dioxide can lead to lung diseases.

Sources of Air Pollutants

The steps are to be taken to control air pollution at source and after the release of pollutants.

(1) *Automobiles Air Pollution* : The following are some of the measure to be followed for controlling air pollution by automobiles.

  (a) To check pollutants smission from vehicular exhaust,
  (b) To chontrol evaporation from fuel tank and carburettor,
  (c) To use the filters, and
  (d) To control through law—Motor Vehicles Act 1988 and other Acts of engine.

(2) *Industrial Air Pollution* : Air pollution by industry and power plants waste can be checked by devising measures for removal of the particulate matter and gaseous pollutants from the wastes. The three types of equipment or devices used are—

  (a) Cyclone collectors,

(b) Electrostatic precipitators, and

(c) Control through law—Acts of industry.

(3) *Gaseous Pollutants* : The following four methods can be used to control air pollution from gaseous pollutants:

(a) *Wet systems :* The alkali liquid reacts with sulphur dioxide to produce a precipitate.

(b) *Dry systems :* Water is in contact with sulphur dioxide produces sulphuric acid.

(c) *Wet dry systems :* This method is very effective in dry cleaning plants, printing shops, paint factories, food processing plants, etc.

(d) *Control through Law :* Acts for Industries are to be enforced to prevent gaseous pollution.

Effects of Air Pollutants on Environment

Like photochemical pollutants, sulfur oxides contribute to the incidence of respiratory diseases. Acid rain, a form of precipitation that contains high levels of sulfuric or nitric acids, can contaminate drinking water and vegetation, damage aquatic life, and erode buildings. When a weather condition known as a temperature inversion prevents dispersal of smog, inhabitants of the area, especially children and the elderly and chronically ill, are warned to stay indoors and avoid physical stress. The dramatic and debilitating effects of severe air pollution episodes in cities throughout the world—such as the London smog of 1952 that resulted in 4,000 deaths—have alerted governments to the necessity for crisis procedures. Even everyday levels of air pollution may insidiously affect health and behaviour. Indoor air pollution is a problem in developed countries, where efficient insulation keeps pollutants inside the structure. In less developed nations, the lack of running water and indoor sanitation can encourage respiratory infections. Carbon monoxide, for example, by driving oxygen out of the bloodstream, causes apathy, fatigue, headache, disorientation, and decreased muscular coordination and visual acuity.

Air pollution may possibly harm populations in ways so subtle or slow that they have not yet been detected. For that reason research is now under way to assess the long-term effects of chronic exposure to low levels of air pollution—what most people experience—as well as to determine how air pollutants interact with one another in the body and with physical factors such as nutrition, stress, alcohol, cigarette smoking, and common medicines. Another subject of investigation is the relation of air pollution to cancer, birth defects, and genetic mutations.

A recently discovered result of air pollution are seasonal "holes" in the ozone layer in the atmosphere above Antarctica and the Arctic, coupled with growing evidence of global ozone depletion. This can increase the amount of ultraviolet radiation reaching the earth, where it damages crops and plants and can lead to skin cancer and cataracts. This depletion has been caused largely by the emission of chlorofluorocarbons (CFCs) from refrigerators, air conditioners, and aerosols.

Vegetation and Controls on Air Pollution

Damage to vegetation by air pollution is of many kinds. Sulfur dioxide may damage field crops such as alfalfa and trees such as pines, especially during the growing season. Both hydrogen fluoride (HF) and nitrogen dioxide (NO2) in high concentrations have been shown to be harmful to citrus trees and ornamental plants, which are of economic, importance in central Florida. Ozone and ethylene are other contaminants that cause damage to certain kinds of vegetation.

Air pollution can affect the dynamics of the atmosphere through changes in long wave and shortwave radiation processes. Particles can absorb or reflect incoming short-wave solar radiation, keeping it from the Earth's surface during the day. Greenhouse gases can absorb long-wave radiation emitted by the Earth's surface and atmosphere.

Carbon dioxide, methane, fluorocarbons, nitrous oxides, ozone, and water vapor are important greenhouse gases. These represent a class of gases that selectively absorb long-wave radiation. This effect warms the temperature of the Earth's atmosphere and surface higher than would be found in the absence of an atmosphere (the greenhouse effect). Because the amount of greenhouse gases in the atmosphere is rising, there is a possibility that the temperature of the atmosphere will gradually rise, possibly resulting in a general warming of the global climate over a time period of several generations.

Green house effect is the phenomenon whereby the earth's atmosphere traps solar radiation, caused by the presence in the atmosphere of gases such as carbon dioxide, water vapour, and methane that allow incoming sunlight to pass through but absorb heat radiated back from the earth's surface.

Controls on Air pollution

There are various air pollution control technologies and land use planning (the term used for a branch of public policy which encompasses various disciplines which seek to order and regulate the use of land in an efficient and ethical way)strategies available to reduce air pollution. At its most basic level land use planning is likely to involve zoning and transport infrastructure planning. In most developed countries, land use planning is an important part of social policy, ensuring that land is used efficiently for the benefit of the wider economy and population as well as to protect the environment.

Efforts to reduce pollution from mobile sources includes primary regulation (many developing countries have permissive regulations), expanding regulation to new sources (such as transport ships, farm equipment, and small gas-powered equipment such as lawn trimmers, chainsaws, and snowmobiles), increased fuel efficiency (such as through the use of hybrid vehicles), conversion to cleaner fuels (such as bioethanol, biodiesel, or conversion to electric vehicles).

Control devices

The following items are commonly used as pollution control devices by industry or transportation devices. They can either destroy contaminants or remove them from an exhaust stream before it is emitted into the atmosphere.

Particulate control

Dust Cyclones- Cyclonic separation is a method of removing

particulates from an air, gas or water stream, without the use of filters, through vortex separation. Rotational effects and gravity are used to separate mixtures of solids and fluids. A high speed rotating (air) flow is established within a cylindrical or conical container called a cyclone. Air flows in a spiral pattern, beginning at the top (wide end) of the cyclone and ending at the bottom (narrow) end before exiting the cyclone in a straight stream through the center of the cyclone and out the top. Larger (denser) particles in the rotating stream have too much inertia to follow the tight curve of the stream and strike the outside wall, falling then to the bottom of the cyclone where they can be removed. In a conical system, as the rotating flow moves towards the narrow end of the cyclone the rotational radius of the stream is reduced, separating smaller and smaller particles. The cyclone geometry, together with flow rate, defines the cut point of the cyclone. This is the size of particle that will be removed from the stream with 50% efficiency. Particles larger than the cut point will be removed with a greater efficiency and smaller particles with a lower efficiency. Large scale cyclones are used in saw mills to remove sawdust from extracted air. Cyclones are also used in oil refineries to separate oils and gases, and in the cement industry.

Electrostatic precipitators An electrostatic precipitator (ESP), or electrostatic air cleaner is a particulate collection device that removes particles from a flowing gas (such as air) using the force of an induced electrostatic charge. Electrostatic precipitators are highly efficient filtration devices that minimally impede the flow of gases through the device, and can easily remove fine particulate matter such as dust and smoke from the air stream.

Baghouses Designed to handle heavy dust loads, a dust collector consists of a blower, dust filter, a filter-cleaning system, and a dust receptacle or dust removal system (distinguished from air cleaners which utilize disposable filters to remove the dust).

Scrubber systems are a diverse group of air pollution control devices that can be used to remove some particulates and/or gases from industrial exhaust streams. Traditionally, the term "scrubber" has referred to pollution control devices that use liquid to wash unwanted pollutants from a gas stream. Recently, the term is also used to describe systems that inject a dry reagent or slurry into a dirty exhaust stream to "wash out" acid gases. Scrubbers are one of the primary devices that control gaseous emissions, especially acid gases. Scrubbers can also be used for heat recovery from hot gases.

Cyclonic spray scrubbers are an air pollution control technology. The cyclonic spray scrubber forces the inlet gas up through the chamber from a bottom tangential entry. Liquid sprayed from nozzles on a center post (manifold) is directed toward the chamber walls and through the swirling gas. This type of technology is a part of the group of air pollution controls collectively referred to as wet scrubbers.

Wet scrubber is a form of pollution control technology. The term describes a variety of devices that remove pollutants from other gas streams. In a wet scrubber, the polluted gas stream is brought into contact with the scrubbing liquid, by spraying it with the liquid, by forcing it through a pool of liquid, or by some other contact method, so as to remove the pollutants. The design of wet scrubbers or any air pollution control device depends on the industrial process conditions and the nature of the air pollutants involved. Inlet gas characteristics and dust properties (if particles are present) are of primary importance. Scrubbers can be designed to collect particulate matter and/or gaseous pollutants. Wet scrubbers remove dust particles by capturing them in liquid droplets. Wet scrubbers remove pollutant gases by dissolving or absorbing them into the liquid.

A wet scrubber's ability to collect small particles is often directly proportional to the power input into the scrubber. Low energy devices such as spray towers are used to collect particles larger than 5 micrometers. To obtain high efficiency removal of 1 micrometer (or less) particles generally requires high energy devices such as venturi scrubbers or augmented devices such as condensation scrubbers. Additionally, a properly designed and operated entrainment separator or mist eliminator is important to achieve high removal efficiencies. The greater the number of liquid droplets that are not captured by the mist eliminator the higher the potential emission levels.

Wet scrubbers that remove gaseous pollutants are referred to as absorbers. Good gas-to-liquid contact is essential to obtain high removal efficiencies in absorbers. A number of wet scrubber designs are used to remove gaseous pollutants, with the packed tower and the plate tower being the most common.

If the gas stream contains both particle matter and gases, wet scrubbers are generally the only single air pollution control device that can remove both pollutants. Wet scrubbers can achieve high removal efficiencies for either particles or gases and, in some instances, can achieve high removal efficiency for both pollutants in the same system. However, in many cases, the best operating conditions for particles collection are the poorest for gas removal.

Water Pollution

Water is the most important element in the biosphere because on one hand, it is vital for the maintainence of all forms of life and on the other hand it helps in the movement, circulation and cycline of nutrients in the biosphere. It is found in various phases and in various storages such as *(i)* in liquid phase (storages such as rivers, lakes, seas and oceans, soils, living organisms etc.), *(ii)* in solid phase (storages such as glaciers and ice sheets and ice caps like ice caps of Arctic region, Greenland and Antarctica and mountain glaciers *e. g.* Alpine glaciers, Himalayan glaciers etc.) Water is also essential for power generation, navigation, irrigation of crops, disposal of total amount of water of the hydrosphere is available to human beings and other biotic communities from various sources such as ground water, rivers, lakes, soils, atmosphere and biological system but ground water provides the largest amount of water.

Water may be present in rivers, ponds lakes and in ground reservoirs. The surface water bodies get easily polluted whereas ground water reservoirs become polluted only after prolonged exposure.

Water pollutants come from numerous natura and anthropogenic sources. Likewise, water pollutants produced in one nation may flow into others, creating complex international control problems that may take decades to solve.

The substances which degrade the quality of water are called water pollutants. The water pollutants are created from two basic cources *e. g.*

(i) Natural Sources of water pollutants include soil erosion, landslides, coastal and cliff erosion, volcanic eruption and decay and decomposition of plants and animals. Excessive soil erosion in the catchment area of particular river increases the sediment load of the river and thus increases the turbity of river and lake.

(ii) Anthropogemic Sources—It may be pointed out that natural water system is capable of taking care of natural pollutants and therefore it is the anthropogenic sources of water pollution include industrial source, urban source, agricultural source, cultural source (congregation of large number of people during pilgrimage, religious fairs etc.

Urban source contributes water pollutants such as sewage, huge quantity of municipal and domestic garbages, industrial effluents from the industrial units located in the urban centres, fallout of particulate matter of automobile exhausts etc.

Various types of chemical fertilizers, pesticides and herbicides etc. are the pollutants which are derived from agricultural sources. These chemical substances are brought the rivers and lakes through surface runoff caused by rainfall and are also moved downward by infiltrating rainwater to reach groundwater.

Industrial sources pollute streams, rivers, lakes and coastal waters through industrial effluents, solid and dissolved chemical pollutants and numerous metals. Besides, fallout of radioactive substances is very dangerous source of air and water pollution.

Water pollution is a major problem in the global context. It has been suggested that it is the leading worldwide cause of deaths and diseases, and that it accounts for the deaths of more than 14,000 people daily. In addition to the acute problems of water pollution in developing countries, industrialized countries continue to struggle with pollution problems as well. In the most recent national report on water quality in the United States, 45 percent of assessed stream miles, 47 percent of assessed lake acres, and 32 percent of assessed bay and estuarine square miles were classified as polluted. Water pollution, by the release of waste products and contaminants into surface runoff into river drainage systems, leaching into groundwater, liquid spills, waste water discharges, petrification and littering. Water pollution is the contamination of water bodies such as lakes, rivers, oceans, and groundwater caused by human activities, which can be harmful to organisms and plants that live in these water bodies.

Effects of Water Pollutants on the Living Organisms

The effects of water pollution are not only devastating to people but also to animals, fish, and birds. Polluted water is unsuitable for drinking, recreation, agriculture, and industry. It diminishes the aesthetic quality of lakes and rivers. More seriously, contaminated water destroys aquatic life and reduces its reproductive ability. Eventually, it is a hazard to human health. Nobody can escape the effects of water pollution.

Water Pollution comes from many different sources and can affect many different things. The effects of water pollution are not only devastating to people, but they can kill animals, fish, and birds. Furthermore, the effects of water pollution pose a serious threat to society today and in the future. There are many forms of water pollution which affects the living organisms.

Effects of Run-Off Pollution : When rain runs off the land it picks up dirt and silt and carries them into the water. When the dirt and silt (sediment) settle in the water body they enter, these sediments can keep sunlight from reaching aquatic plants, plants that live in and grow in the water. When the sun can't reach the plants, they die. The sediments can also clog fish gills, and smother organisms that live on the bottom of the body of water.

Contaminated Ground Water Effects : When contaminated water seeps into the ground it can have serious effects. People can get very sick and have the possibility of developing liver or kidney problems, cancer or other illnesses, depending on if contaminated water seeps into the ground.

Effects of Oil Pollution and Antifreeze : When oil is spilled into the water, the effects on the ecosystem and its components are devastating. Some animals, such as birds, mammals, and fish may and can be killed if they ingest oil. Many may die from eating oil contaminated prey. Birds may die if the oil coats their feathers. They can neither fly nor stay warm. Furthermore, when oil coats the feathers they can become sick and die. Oil and antifreeze can cause the water to have a bad odor and leave a sticky film on the surface of water that kills animals and fish. Oil is one of the most devastating pollutants of water.

Fertilizers and Other Chemicals : Water, from rain, runs down the slopes of the land which may include farm areas that use fertilizers, pesticides, and other farming chemicals. After that they travel down into the rivers, lakes, or oceans. Fertilizers and some chemicals may cause plants to grow quicker. With the growth of more plants, more bacteria will grow (bacteria eat dead plants). Bacteria need oxygen to survive and if there are more bacteria in a river than normal, there is less oxygen for fish and some of them may die. In addition, some fish in the Great Lakes suffer from tumors and the populations of some species in these waters are declining. Other chemicals besides farming chemicals effect humans as well. Nitrates in drinking water can cause diseases to infants that might cause them death. Cadmium (a metal in sludge-derived fertilizer) can be absorbed by crops, and if people ingest this in sufficiently it can cause diarrheal disorders, liver, and kidney damage. The culprit is suspected to be inorganic substances such as mercury, arsenic, and lead. Also, other chemicals can cause problems with the taste, smell, and the color of water. Pesticides, PCB's, and PCP's (polychlorinated phenols) are some examples that are toxic to all life. Pesticides are used in farming, forestry, and homes. PCB's can still be found as insulators in old electrical transformers, and PCP's can be found in products such as wood preservatives. They are very toxic and that is what makes them a threat to our ecosystem.

Effects of Factory Pollution : Many factories have pipes that drain chemicals into rivers or streams. These chemicals can damage aquatic life as they are carried downstream. Furthermore, the added chemicals can warm the river, which decreases the amount of oxygen that the fish need to live.

Effects of Garbage from Private Offices and Homes : Many people today dump their garbage into streams, lakes, rivers, and oceans. Some examples of this garbage are cans, paper, furniture, and other household products. When people dump cleaning products into the ecosystem they are endangering its inhabitants. When plastic is dumped in lakes, ducks are at risk because they might be strangled and when dumped in the ocean, dolphins might be killed. Aluminum cans can cut the animals and fish.

Effects of Eutrophication : Eutrophication, occurs when lake water is artificially supplemented with nutrients, which causes abnormal plant growth. The cause of eutrophication can be runoff of chemical fertilizers from fields. Eutrophication can produce problems such as bad tastes and odors as well as green scum algae. Also, the growth of rooted plants increases, which decreases the amount of oxygen in the deepest waters of the lake. A common chemical change is the precipitation of calcium carbonate in hard waters. Eutrophication makes some lakes void of life.

Effects of Acid Rain : The effects of acid rain are most clearly seen in lakes, streams, rivers, oceans, and other bodies of water. Acid rain directly falls on water, but it can flow into rivers after it falls on land. Lakes and streams become acidic (pH value goes down) when the water and the land around it can not neutralize the acid rain. Animals that live in the water environment are hurt and possibly killed. Some fish can only tolerate a certain amount of acid before dying. The more acid rain that falls, the life in the bodies of water decreases. Furthermore, animals that eat prey that is affected will be killed because they will be consuming acid.

Controls on Water Pollution

We all thrive on clean water. It is the source of all life, without it life would be non-existent. That is why we all need to do our part protecting. Controlling water pollution is our only hope of keeping the waters of the world clean.

<u>Around the House</u>

- Recycle reusable objects
- Dispose of chemicals properly
- Dispose of pet waste in garbage or toilet

- Do not use pesticides, or fertilizers frequently or before a heavy rain
- Use absorbent materials to clean spills from surfaces
- Rain spouts, sprinkler systems, and garden hoses should wash off on to grassy areas instead of paved surfaces
- Keep litter, pet wastes, leaves, and debris out of street gutters and storm drains
- Plant a ground cover and stabilize erosion prone areas to control soil erosion
- Encourage your town to develop ordinances on controlling water pollution

<u>Automobiles</u>

- Wash cars at commercial car washes, or wash them without soap over grass.
- Make sure your car does not leak oil, antifreeze and other fluids.
- Dispose of used motor oil and antifreeze at a service station with a recycling center or to a toxic waste center.

<u>Agriculture</u>

- Manage animal waste
- Reduce soil erosion by using processes like no-till, contour farming, and strip cropping.
- Use planned grazing procedures.
- Dispose of pesticides, containers, and tank rinsate in a an approved manner.

Measures to prevent water pollution essentially strive to conserve and protect water quality - in terms of its use reduction and disposal, waste water treatment, procedural changes and recycling. Water conservation programmes have also included the substitution/ reduction of hazardous materials and the generation of hazardous wastes. Employee awareness, education and training in pollution identification and reduction is critical in achieving successful results.

Sewage treatment : Sewage treatment, or domestic wastewater treatment, is the process of removing contaminants from wastewater and household sewage, both runoff and domestic. It includes physical, chemical and biological processes to remove physical, chemical and biological contaminants. Its objective is to produce a waste stream and a solid waste suitable for discharge or reuse back into the environment. This material is often contaminated with many toxic organic and inorganic compounds.

Sewage is created by residences, institutions, hospitals and commercial and industrial establishments. It can be treated close to where it is created or collected and transported via a network of pipes and pump stations to a municipal treatment plant. Sewage collection and treatment is typically subject to local, state and federal regulations and standards. Industrial sources of wastewater often require specialized treatment processes.

The sewage treatment involves three stages, called primary, secondary and tertiary treatment. First, the solids are separated from the wastewater stream. Then dissolved biological matter is progressively converted into a solid mass by using water-borne micro-organisms. Finally, the biological solids are neutralized then disposed of or re-used, and the treated water may be disinfected chemically or physically. The final effluent can be discharged into a stream, river, bay, lagoon or wetland, or it can be used for the irrigation of a golf course, green way or park. If it is sufficiently clean, it can also be used for groundwater recharge or agricultural purposes.

Industrial wastewater treatment : Industrial wastewater treatment covers the mechanisms and processes used to treat waters that have been contaminated in some way by anthropogenic industrial or commercial activities prior to its release into the environment or its reuse.

Most industries produce some wet waste although recent trends in the developed world have been to minimise such production or recycle such waste within the production process. However, many industries remain dependent on processes that produce wastewater.

Treatment of industrial wastewater : The different types of contamination of wastewater require a variety of strategies to remove the contamination.

Solids removal : Most solids can be removed using simple sedimentation techniques with the solids recovered as slurry or sludge. Very fine solids and solids with densities close to the density of water pose special problems. In such case filtration or ultrafiltration may be required.

Oils and grease removal : Many oils can be recovered from open water surfaces by skimming devices. However, hydraulic oils and the majority of oils that have degraded to any extent will also have a soluble or emulsified component that will require further treatment to eliminate. Dissolving or emulsifying oil using surfactants or solvents usually exacerbates the problem rather than solving it, producing wastewater that is more difficult to treat.

The wastewaters from large-scale industries such as oil refineries, petrochemical plants, chemical plants, and natural gas processing plants commonly contain gross amounts of oil and suspended solids. Those industries use a device known as an API oil-water separator which is designed to separate the oil and suspended solids from their wastewater effluents. The name is derived from the fact that such separators are designed according to standards published by the American Petroleum Institute (API).

The API separator is a gravity separation device designed by using Stokes Law to define the rise velocity of oil droplets based on their density and size. The design is based on the specific gravity difference between the oil and the wastewater because that difference is much smaller than the specific gravity difference between the suspended solids and water. The suspended solids settles to the bottom of the separator as a sediment layer, the oil rises to top of the separator and the cleansed wastewater is the middle layer between the oil layer and the solids.

Typically, the oil layer is skimmed off and subsequently re-processed or disposed of, and the bottom sediment layer is removed by a chain and flight scraper (or similar device) and a sludge pump. The water layer is sent to further treatment consisting usually of a dissolved air flotation (DAF) unit for additional removal of any residual oil and then to some type of biological treatment unit for removal of undesirable dissolved chemical compounds.

Parallel plate separators are similar to API separators but they include tilted parallel plate assemblies (also known as parallel packs). The parallel plates provide more surface for suspended oil droplets to coalesce into larger globules. Such separators still depend upon the specific gravity between the suspended oil and the water. However, the parallel plates enhance the degree of oil-water separation. The result is that a parallel plate separator requires significantly less space than a conventional API separator to achieve the same degree of separation.

Removal of biodegradable organics : Biodegradable organic material of plant or animal origin is usually possible to treat using extended conventional wastewater treatment processes such as activated sludge or filter. Problems can arise if the wastewater is excessively diluted with washing water or is highly concentrated such as neat blood or milk. The presence of cleaning agents, disinfectants, pesticides, or antibiotics can have detrimental impacts on treatment processes.

Activated sludge process : Activated sludge is a biochemical process for treating sewage and industrial wastewater that uses air (or oxygen) and microorganisms to biologically oxidize organic pollutants, producing a waste sludge containing the oxidized material. In general, an activated sludge process includes:

An aeration tank where air (or oxygen) is injected and thoroughly mixed into the wastewater.

A settling tank (usually referred to as a "clarifier" or "settler") to allow the waste sludge to settle. Part of the waste sludge is recycled to the aeration tank and the remaining waste sludge is removed for further treatment and ultimate disposal.

Trickling filter process : A trickling filter consists of a bed of rocks, gravel, slag, peat moss, or plastic media over which wastewater flows downward and contacts a layer (or film) of microbial slime covering the bed media. Aerobic conditions are maintained by forced air flowing through the bed or by natural convection of air. The process involves adsorption of organic compounds in the wastewater by the microbial slime layer, diffusion of air into the slime layer to provide the oxygen required for the biochemical oxidation of the organic compounds. The end products include carbon dioxide gas, water and other products of the oxidation. As the slime layer thickens, it becomes difficult for the air to penetrate the layer and an inner anaerobic layer is formed.

The components of a complete trickling filter system are: fundamental components:

A bed of filter medium upon which a layer of microbial slime is promoted and developed.

An enclosure or a container which houses the bed of filter medium.

A system for distributing the flow of wastewater over the filter medium.

A system for removing and disposing of any sludge from the treated effluent.

The treatment of sewage or other wastewater with trickling filters is among the oldest and most well characterized treatment technologies.A trickling filter is also often called a trickle filter, trickling biofilter, biofilter, biological filter or biological trickling filter.

Treatment of other organics : Synthetic organic materials including solvents, paints, pharmaceuticals, pesticides, coking products and so forth can be very difficult to treat. Treatment methods are often specific to the material being treated. Methods include Advanced Oxidation Processing, distillation, adsorption, vitrification, incineration, chemical immobilisation or landfill disposal. Some materials such as some detergents may be capable of biological degradation and in such cases, a modified form of wastewater treatment can be used.

Treatment of acids and alkalis : Acids and alkalis can usually be neutralised under controlled conditions. Neutralisation frequently produces a precipitate that will require treatment as a solid residue that may also be toxic. In some cases, gasses may be evolved requiring treatment for the gas stream. Some other forms of treatment are usually required following neutralisation.

Waste streams rich in hardness ions as from de-ionization processes can readily lose the hardness ions in a buildup of precipitated calcium and magnesium salts. This precipitation process can cause severe furring of pipes and can, in extreme cases, cause the blockage of disposal pipes. A 1 metre diameter industrial marine discharge pipe serving a major chemicals complex was blocked by such salts in the 1970s. Treatment is by concentration of de-ionization waste waters and disposal to landfill or by careful pH management of the released wastewater.

Treatment of toxic materials : Toxic materials including many organic materials, metals (such as zinc, silver, cadmium, thallium, etc.) acids, alkalis, non-metallic elements (such as arsenic or selenium) are generally resistant to biological processes unless very dilute. Metals can often be precipitated out by changing the pH or by treatment with other chemicals. Many, however, are resistant to treatment or mitigation and may require concentration followed by landfilling or recycling. Dissolved organics can be incinerated within the wastewater by Advanced Oxidation Processes.

Agricultural wastewater treatment : Agricultural wastewater treatment relates to the treatment of wastewater produced in the course of agricultural activities. As agriculture is a highly intensified industry in many parts of the world, the range of wastewaters requiring treatment can encompass at least the following:

Animals wastes - both liquid and solid
Silage liquor
Piggery waste
Pesticide run off and surpluses
Milking parlour wastes including milk
Slaughtering waste
Vegetable washing water
Fire water

Animal wastes both Liguid and Solid

Animal wastes from cattle can be as produced as solid or semisolid manure or as a liquid slurry. The production of slurry is especially common in housed dairy cattle.

Treatment : Whilst solid manure heaps outdoors can give rise to polluting wastewaters from rain washing, this type of waste is usually relatively easy to treat by containment and/or covering of the heap.

Animal slurries require special handling and are usually treated by containment in lagoons before disposal by spray or trickle application to grassland. Constructed wetlands are sometimes used to facilitate treatment of animal wastes, as are anaerobic lagoons. Excessive application or application to sodden land or insufficient land area can result in direct runoff to watercourses with the potential for causing severe pollution. Application of slurries to land overlying aquifers can result in direct contamination or, more commonly, elevation of nitrogen levels as nitrite or nitrate.

The disposal of any wastewater containing animal waste upstream of a drinking water intake can pose serious health problems to those drinking the water because of the highly resistant spores present in many animals that are capable of causing disabling in humans. This risk exists even for very low level seepage via shallow surface drains or from rainfall run-off.

Some animal slurries are treated by mixing with straws and composted at high temperature to produce a bacteriologically sterile and friable manure for soil improvement.

Piggery waste : Piggery waste is comparable to other animal wastes except that many piggery wastes contain elevated levels of copper that can be toxic in the natural environment. Ascarid worms and their eggs are also common and can infect humans if wastewater treatment is ineffective.

Treatment : As for general animal waste although the liquid fraction of the waste is frequently separated off and re-used in the piggery to avoid the prohibitively expensive costs of disposing of a copper rich liquor.

Silage liquor : Fresh or wilted grass or other green crops can be made into the semi fermented product called silage which can be stored and used as winter forage for cattle and sheep. The production of silage often involves the use of an acid conditioner such as sulfuric acid or formic acid. The process of silage making frequently produces a yellow-brown strongly smelling liquid which is very rich in simple sugars, alcohol, short-chain organic acids and silage conditioner. This liquor is one of the most polluting organic substances known. The volume of silage liquor produced is generally in proportion to the moisture content of the ensiled material.

Treatment : Silage liquor is best treated through prevention by wilting crops well before silage making. Any silage liquor that is produced can be used as part of the food for pigs. The most effective treatment is by containment in a slurry lagoon and subsequently spread on land following substantial dilution with slurry. Containment of silage liquor on its own can cause structural problems in concrete pits because of the acidic nature of silage liquor.

Pesticide runoff and surpluses : Inappropriate use of pesticides so that pesticide-containing wastewaters enter the environment can give rise to severe and long-lasting ecological damage. This is particularly true for insecticides used in sheep dips because of the volumes of pesticide-containing wastewater requiring disposal and because of the persistent and damaging nature of the pesticides.

Treatment : There are few safe ways of disposing of pesticide surpluses other than through containment in well managed landfills or by incineration. In some parts of the world, spraying on land is a permitted method of disposal.

Milking parlour wastes including milk : Although milk has a deserved reputation as an important and valuable food product, its presence in wastewaters is highly polluting because of its organic strength, which can lead to very rapid de-oxygenation of receiving waters. Milking parlour wastes also contain large volumes of wash-down water, some animal waste together with cleaning and disinfection chemicals.

Treatment : Milking parlour waste is often treated in admixture with human sewage in a local sewage treatment plant. This ensures that disinfectants and cleaning agents are sufficiently diluted and amenable to treatment. Running milking wastewaters into a farm slurry lagoon is a possible option although this tends to consume lagoon capacity very quickly. Land spreading is also a treatment option.

Slaughtering waste : Wastewater from slaughtering activities is similar to milking parlour waste (see above) although considerably stronger in its organic composition and therefore potentially much more polluting.

Treatment : As for milking parlour waste (see above).

Vegetable washing water : Washing of vegetables produces large volumes of water contaminated by soil and vegetable pieces. Low levels of pesticides used to treat the vegetables may also be present together with moderate levels of disinfectants such as chlorine.

Treatment : Most vegetable washing waters are extensively recycled with the solids removed by settlement and filtration. The recovered soil can be returned to the land

Fire water : Although few farms plan for fires, fires are nevertheless more common on farms than on many other industrial premises. Stores of pesticides, herbicides, fuel oil for farm machinery and fertilisers can all help promote fire and can all be present in environmentally lethal quantities in wastewater from firefighting at farms.

Treatment : All farm environmental management plans should allow for containment of substantial quantities of firewater and for its subsequent recovery and disposal by specialist disposal companies. The concentration and mixture of contaminants in fire-water make them unsuited to any treatment method available on the farm. Even land spreading has produced severe taste and odour Problems For Downstream Water Supply Companies In The Past.

Soil Pollution

Soils are infact the very heart of the life layer (the biosphere) because these represent a zone wherein plant nutrients are produced, held, maintained and are made available to plants through their roots and to the micro-organisms which live in the soils, Soil is also very important environmental attribute for human society because (i) it is the basic medium for food and timber production; (ii) it provides formation for building and roads; (iii) it is very important exhaustible natural resource because it cannot be replaced if it is destroyed or lost through excessive soil erosion caused by atmospheric activities.

The formation of soil is a slow process as the formation and development of one inch of soil require about one thousand years whereas the destruction of soils through erosion and pollution is quick process, the quality of soils depends upon the nutrients (both organic and inorganic), humus content, moisture, temperature etc. present in the soils.

Environmental World Document

Soil Pollution : Soil contamination is caused by the presence of man-made chemicals or other alteration in the natural soil environment. This type of contamination typically arises from the rupture of underground storage tanks, application of pesticides, percolation of contaminated surface water to subsurface strata, oil and fuel dumping, leaching of wastes from landfills or direct discharge of industrial wastes to the soil. The most common chemicals involved are petroleum hydrocarbons, solvents, pesticides, lead and other heavy metals. This occurrence of this phenomenon is correlated with the degree of industrialization and intensity of chemical usage.

The concern over soil contamination stems primarily from health risks, both of direct contact and from secondary contamination of water supplies. Mapping of contaminated soil sites and the resulting cleanup are time consuming and expensive tasks, requiring extensive amounts of geology, hydrology, chemistry and computer modelling skills.

Soil pollution is defined as the build-up in soils of persistent toxic compounds, chemicals, salts, radioactive materials, or disease causing agents, which have adverse effects on plant growth and animal health.

Soil pollution comprises the pollution of soils with materials, mostly chemicals that are out of place or are present at concentrations higher than normal which may have adverse effects on humans or other organisms. It is difficult to define soil pollution exactly because different opinions exist on how to characterize a pollutant; while some consider the use of pesticides acceptable if their effect does not exceed the intended result, others do not consider any use of pesticides or even chemical fertilizers acceptable. However, soil pollution is also caused by means other than the direct addition of xenobiotic (man-made) chemicals such as agricultural runoff waters, industrial waste materials, acidic precipitates, and radioactive fallout.

Both organic (those that contain carbon) and inorganic (those that don't) contaminants are important in soil. The most prominent chemical groups of organic contaminants are fuel hydrocarbons, polynuclear aromatic hydrocarbons (PAHs), polychlorinated biphenyls (PCBs), chlorinated aromatic compounds, detergents, and pesticides. Inorganic species include nitrates, phosphates, and heavy metals such as cadmium, chromium and lead; inorganic acids; and radionuclides (radioactive substances). Among the sources of these contaminants are agricultural runoffs, acidic precipitates, industrial waste materials, and radioactive fallout.

Soil pollution can lead to water pollution if toxic chemicals leach into groundwater, or if contaminated runoff reaches streams, lakes, or oceans. Soil also naturally contributes to air pollution by releasing volatile compounds into the atmosphere. Nitrogen escapes through ammonia volatilization and denitrification. The decomposition of organic materials in soil can release sulfur dioxide and other sulfur compounds, causing acid rain. Heavy metals and other potentially toxic elements are the most serious soil pollutants in sewage. Sewage sludge contains heavy metals and, if applied repeatedly or in large amounts, the treated soil may accumulate heavy metals and consequently become unable to even support plant life.

In addition, chemicals that are not water soluble contaminate plants that grow on polluted soils, and they also tend to accumulate increasingly toward the top of the food chain. The banning of the pesticide DDT in the United States resulted from its tendency to become more and more concentrated as it moved from soil to worms or fish, and then to birds and their eggs. This occurred as creatures higher on the food chain ingested animals that were already contaminated with the pesticide from eating plants and other lower animals. The ever-increasing pollution of the environment has been one of the greatest concerns for science and the general public in the last fifty years. The rapid industrialization of agriculture, expansion of the chemical industry, and the need to generate cheap forms of energy has caused the continuous release of man-made organic chemicals into natural ecosystems. Consequently, the atmosphere, bodies of water, and many soil environments have become polluted by a large variety of toxic compounds. Many of these compounds at high concentrations or following prolonged exposure have the potential to produce adverse effects in humans and other organisms: These include the danger of acute toxicity, mutagenesis (genetic changes), carcinogenesis, and teratogenesis (birth defects) for humans and other organisms. Some of these man-made toxic compounds are also resistant to physical, chemical, or biological degradation and thus represent an environmental burden of considerable magnitude.

Numerous attempts are being made to decontaminate polluted soils, including an array of both in situ (on-site, in the soil) and off-site (removal of contaminated soil for treatment) techniques. None of these is ideal for remediating contaminated soils, and often, more than one of the techniques may be necessary to optimize the cleanup effort.

The most common decontamination method for polluted soils is to remove the soil and deposit it in landfills or to incinerate it. These methods, however, often exchange one problem for another: landfilling merely confines the polluted soil while doing little to decontaminate it, and incineration removes toxic organic chemicals from the soil, but subsequently releases them into the air, in the process causing air pollution.

For the removal and recovery of heavy metals various soil washing techniques have been developed including physical methods, such as attrition scrubbing and wet-screening, and chemical methods consisting of treatments with organic and inorganic acids, bases, salts and chelating agents.

"The contamination of soil with excess of chemicals, fertilizers, insecticides, herbicides is known as soil pollution."

The decrease in the quality of soils either due to human activities or natural sources or by both is known as soil pollution or soil degradation. The soil pollution is caused due to soil erosion, decrease in plant nutrients, decrease in soil micro organisme, excess or deficit of moisture content, high fluctuation of temperature and lack of humas content.

Major Soil Pollutants and their Sources: Major soil pollutants are fluorides, nitrogen fertilization, weedicides, heavy metals and their salts:

1. Fluorides : They combine with Mg2+ of chlorophyll and hence inhibit photosynthesis, cause leaf abscission and of fruit. Maize is the sensitive indicator of fluoride pollution. In human beings mottling of teeth (fluorosis) is an indication of fluorination. Bone fluorosis results in weak bones, boat shaped posture and knocking of knee.

2. Nitrogen fertilization (nitrates + nitrite) : Toxic concentration in leaves and fruits enters into food chain. In alimentary canal, activity of bacteria changes nitrates into nitrites. The latter enters blood and combine with haemoglobin to form met6-haemoglobin so oxygen transport is reduced. It gives rise to disease called methanaemoglobinaemia. In infants it cause cyanosis (blue babies due to bluish tint of skin). Nitrate poisoning is fatal unless methylene blue is injected (in infants) while in adult it produces breathlessness.

3. Weedicides : They are usually metabolic inhibitors which stop photosynthesis and other metabolic activities killing the plant. Some causes death due to proliferation of phloem cells to block transport of organic food.

Heavy Metals and Their Salts

Natural concentrations of heavy metals in soils depend primarily on the type and chemistry of the parent materials from which the soils are dervied. However, anthropogenic inputs may lead to concentrations highly exceeding those from natural sources. Average concentrations of some heavy metals in the Earth's crust, in some sediments and generally in soils.

We can conclue that lead, cadmium, tin, and mercury, are the most abundant metallic polluants introduced into soil by anthropogenic activities. The means concentration of cadmium in soils is six times its means concentration in the crust. Concentrations of lead, mercury, and tin in soils attain double their mean values in the Earth's crust.

Lead is one of the oldest metallic pollutants introduced by man into the environment. It was called plumbum nigrum by the Romans to differentiate it from plumbum candidum-tin. Water drainage pipes made fo lead and carrying the insignia of some Roman emperors are still in use, despite the fact that they introduce lead as a dangerous pollutant into the soil when they become porous.

The use of lead to make water pipes is no longer allowed due to their dangerous effect on the environment and the harm they cause for humans. Lead can, if ingested with food or water, cause severe damage to the nervous system, the urinary system and the reproductive system. It may cause abortion in females and reduce the fertility of males by severely lowering their capability of sperm production. It may also cause anaemia by obstructing the biosynthesis of haemoglobin.

Cadmium-a heavy metal, which is mainly used in the production of nickel/cadmium batteries or that of pigments and stabilisers for PVC; in metallurgical and electronic industries, it is one of the most frequently registered metalllic pollutants. It passes into the enviroment through emissions in the metallurgical industries or application of phosphate fertilisers in agriculture. Detergents and petroleum products may also contribute to its circulation as an environmental pollutant.

Daily intake of cadmium for humans is estimated to be somewhere between 0.15 and 1.5 μg. Smokers may risk a considerable increase; for every cigarette, about 2 to 4μg cadmium is additionally inhaled.

(2) Socrces of Soil Pollution : The main factors of soil pollution are accelerated rate of soil erosion consequent upon major land use changes (e.g., deforestantion); excessive use of chemical fertilizers, pesticides, insecticide and herbicides; polluted waste water from industrial and urban areas; a few micro-organism; forest fires; industrial solid wastes; water logging and related capillary process; leaching processes; drought etc. Following are the source of soil pollution.

1. *The Physical Source* of soil pollution is related to soil erosion and consequent soil degradation caused by natural and anthropogenic factors. The natural factors of soil erosion include a mount and intensity of rainfall, temperature and wind; vegetation and soil characteristics. These factors are further accelerated by human activities such as land use changes (*e.g.* deforestation). In most of the developing countries of the tropical and subtropical regions accelerated rate of soil erosion due to rapid rate of deforestation and faulty agricultural practices has degraded the

soil on a largescale because the top fertile layer has been washed out.

2- *The Biological Sources of Agents* of soil pollution include those micro-organisms and unwanted plants which degrade the quality and therefore fertility of the soils. The soil pollutant biological agents are grouped into four major categories. *(i)* Pathogenic microorganisms excreted by human beings; *(ii)* Pathogenic microorganisms excreted by domestic animals; *(iii)* Pathogenic microorganisms already present in the soils; and *(iv)* Enteric bacteria and protozoa. These micro-organisms enter the soils from various sources and degrade them. These micro-organisms also enter the food chains and thus affect human bodies.

3. *Air-born Sources* of pollutants are infact are pollutants which are released into the atmosphere by humn volcanoes, (chimneys of factories), automobiles, thermal power plants and domestic sources. In fact, gaseous and solid particulate pollutants emitted by factory chimneys and other sources are transported.

The fallouts of these pollutants are deposited in the soils which are polluted due to toxic substances. Sulphur emitted from the from the factories causes acid rains which lower the pH of the soils. Acid rains increase the acidity of the soils. Highly acidic soils are injurious to plant growth.

Huge quantities of particulate matters emitted from cement factories, lime kilns, coal mining, loading and unloading of coal, thermal power plants etc. reach the soils and thus pollute them by altering their chemical and physical properties.

Flakes and chips of mica from the mica mines spread over the soils and pollute them.

Magnesite dusts, when mixed with soils, cause marked rise in the soil *pH* increase in the soil alkalinity which causes salinization) and decrease in potash (K), calcium (Ca), magnesium (Mg) and phosphorous to critical lower limit. Fallouts of metallic particulate matter from metal smelters into soils damage their physical and chemical properties.

The fallouts of mercury released from industrial processes and combustion of fossil fuels when mixed in he soild reach the food chain. Similarly, high concentration of lead pollutes the soils to large extent.

4. *Chemical Fertilizers and Biocides* Excessive use of chemical fertilizers to boost agricultural production causes alteration in the physical and chemical properties of soils, though chemical fertilizaers act as inorganic plant nutrients. The most dangerous pollutants are different kinds of biocides which destory micro-organisms including useful bacteria. Infact, rapid growth of population and commercialization of agriculture have been responsible for the phenomenal increase in the production and consumption of chemical fertilizers and biocides.

The following types of biocides (pesticides, insecticides, herbicides etc.) are generally used to get rid of unwanted plants and to kill harmful insects and destory pests to boost agricultual production.

5. *Organic Phosphate Cmpounds* e.g. Malathions. These are used to kill insects by damaging their nerve systems. Frequent use of these chemicals results in the accumulation of acetylchlorine in the soils and ultimately these chmicals are transported to plant tissues. The mammals depending on these plants are thus adversely affected. *Chlorinated Hydrocarbons e.g.* D.D.T. dieldrin, aldrin etc. These insecticides are generally used to kill insects and micro-organisms.

6. *Arsenic Containing Pesticides* reach the soils and are transported to plants and thus to foodgrains. These chemicals cause gastric digestive problems in humans. *Sodium Fluoroaccetates are* generally used to kill rodents and accumulate in the soils to reach the food chains.

Improper disposal of industrial and urban wastes and irrigation of agricultural fields from plluted urban sewage water near urban and industrial areas degrade the soil properties by changing their physical and chemical properties.

Effects of Soil Pollution

The effects of soil pollution are: Not enough oxygen in the soil, acidy soils that may burn the plant, bugs will go and start an infestation in the soil, it effects plants growth, the soil pollution eats away at the nutrients and becomes a bigger soil polluter, not enough drainage, new soil diseases develop every year, not enough moisture in the soil. These are the effects of soil pollution:

- pollution runs off into rivers and kills the fish, plants and other aquatic life
- crops and fodder grown on polluted soil may pass the pollutants on to the consumers
- polluted soil may no longer grow crops and fodder
- Soil structure is damaged (clay ionic structure impaired)
- corrosion of foundations and pipelines
- impairs soil stability
- may release vapours and hydrocarbon into buildings and cellars
- may create toxic dusts
- Causes cancers including leukemia
- Lead in soil is especially hazardous for young children causing developmental damage to the brain
- Mercury can increase the risk of kidney damage; cyclodienes can lead to liver toxicity
- Causes neuromuscular blockage as well as depression of the central nervous system
- Also causes headaches, nausea, fatigue, eye irritation and skin rash
- Contact with contaminated soil may be direct (from using parks, schools etc) or indirect (by inhaling soil contaminants which have vaporized)
- Soil pollution may also result from secondary contamination of water supplies and from deposition of air contaminants (for example, via acid rain)
- Contamination of crops grown in polluted soil brings up problems with food security
- Since it is closely linked to water pollution, many effects of soil contamination appear to be similar to the ones caused by water contamination

Effects of Soil Pollution on Animals :

· Can alter metabolism of micro-organisms and arthropods in a given soil environment; this may destroy some layers of the primary food chain, and thus have a negative effect on predator animal species

Small life forms may consume harmful chemicals which may then be passed up the food chain to larger animals; this may lead to increased mortality rates and even animal extinction

Effects of Soil Pollution on Trees and Plants :

- May alter plant metabolism and reduce crop yields
- Trees and plants may absorb soil contaminants and pass them up the food chain

Controls on Soil Pollution

Soil pollution has been slightly controlled by putting regulations on the use of DDT and introduction of alternatives to it. However the task of eliminating completely soil pollution is not easy, third some third world countries still utilize pollutants such as DDT as pesticides. Mining cannot be stopped because we are in constant need for mineral ores for different applications.

Control of Soil Pollution :

1. Use of pesticides should be minimized.
2. Use of fertilisers should be judicious.
3. Cropping techniques should be improved to prevent growth of weeds.
4. Special pits should be selected for dumping wastes.
5. Controlled grazing and forest management.
6. Wind breaks and wind shield in areas exposed to wind erosion
7. Planning of soil binding grasses along banks and slopes prone to rapid erosion.
8. Afforestation and reforestation.

It is very essential to control and prevent soil pollution because the existence of man, animals and vegetation depend upon the quality of soil. It is necessary to maintain and also increase the quality of soils. The essential commodities required for man, animals and plants come from soils. The following are suggested to control soil pollution :

1. To check the soil erosion by using controlling measures.
2. To use judiciously chemical fertilizers and pesticides and insecticides.

3. To restrict the use of D.D.T.
4. To dispose properly urban and industrial wastes.
5. Crop management and proper land use.
6. To educate the farmers about the use of fertilizers and biocides.

Noise Pollution

Noise pollution (or environmental noise) is displeasing human- or machine-created sound that disrupts the activity or balance of human or animal life. A common form of noise pollution is from transportation, principally motor vehicles. The word "noise" comes from the Latin word "noxia" meaning "injury" or "hurt".

The source of most noise worldwide is transportation systems, motor vehicle noise, but also including aircraft noise and rail noise. Poor urban planning may give rise to noise pollution, since side-by-side industrial and residential buildings can result in noise pollution in the residential area.

Other sources are car alarms, office equipment, factory machinery, construction work, grounds keeping equipment, barking dogs, appliances, power tools, lighting hum, audio entertainment systems, loudspeakers and noisy people.

Noise pollution is unwanted human-created sound that disrupts the environment. The dominant form of noise pollution is from transportation sources, principally motor vehicles, referred to as environmental noise. The word noise comes from the Latin word nausea meaning seasickness.

Noise Pollution : Noise is defined as 'unwanted sounds' being 'dumped' into atmosphere to disturb the unwilling ears. It effects our physiological and mental health. A sound of over 115 decides is harmful for ears. They city noise is often sufficient to deafen people gradually, at least partially with advancing age.

Sources of Noise Pollution : High intensity sound from industrial machines, supersonic aeroplanes, bomb blasts, exploding of crackers, blaring radios and loudspeakers, slogans shouting city crowd, traffic noise etc.

Noise pollution like other pollutants is also a by- product of industrialization, urbanizations and modern civilization.

Broadly speaking, the noise pollution has two sources, i.e. industrial and non- industrial. The industrial source includes the noise from various industries and big machines working at a very high speed and high noise intensity. Non- industrial source of noise includes the noise created by transport/vehicular traffic and the neighborhood noise generated by various noise pollution can also be divided in the categories, namely, natural and manmade. Most leading noise sources will fall into the following categories: roads traffic, aircraft, railroads, construction, industry, noise in buildings, and consumer products.

1. Road Traffic Noise : In the city, the main sources of traffic noise are the motors and exhaust system of autos, smaller trucks, buses, and motorcycles. This type of noise can be augmented by narrow streets and tall buildings, which produce a canyon in which traffic noise reverberates.

2. Air Craft Noise : Now-a-days , the problem of low flying military aircraft has added a new dimension to community annoyance, as the nation seeks to improve its nap-of the- earth aircraft operations over national parks, wilderness areas , and other areas previously unaffected by aircraft noise has claimed national attention over recent years.

3. Noise from railroads : The noise from locomotive engines, horns and whistles, and switching and shunting operation in rail yards can impact neighboring communities and railroad workers. For example, rail car retarders can produce a high frequency, high level screech that can reach peak levels of 120 dB at a distance of 100 feet, which translates to levels as high as 138, or 140 dB at the railroad worker's ear.

4. Construction Noise : The noise from the construction of highways, city streets, and buildings is a major contributor to the urban scene. Construction noise sources include pneumatic hammers, air compressors, bulldozers, loaders, dump trucks (and their back-up signals), and pavement breakers.

5. Noise in Industry : Although industrial noise is one of the less prevalent community noise problems, neighbors of noisy manufacturing plants can be disturbed by sources such as fans, motors, and compressors mounted on the outside of buildings Interior noise can also be transmitted to the community through open windows and doors, and even through building walls. These interior noise sources have significant impacts on industrial workers, among whom noise- induced hearing loss is unfortunately common.

6. Noise in building : Apartment dwellers are often annoyed by noise in their homes, especially when the building is not well designed and constructed. In this case, internal building noise from plumbing, boilers, generators, air conditioners, and fans, can be audible and annoying. Improperly insulated walls and ceilings can reveal the sound of-amplified music, voices, footfalls and noisy activities from neighboring units. External noise from emergency vehicles, traffic, refuse collection, and other city noises can be a problem for urban residents, especially when windows are open or insufficiently glazed.

7. Noise from Consumer products : Certain household equipment, such as vacuum cleaners and some kitchen appliances have been and continue to be noisemakers, although their contribution to the daily noise dose is usually not very large.

Harmful Effects : On Human Being, Animal and Property: Noise has always been with the human civilization but it was never so obvious, so intense, so varied & so pervasive as it is seen in the last of this century. Noise pollution makes men more irritable. The effect of noise pollution is multifaceted & inter related. The effects of Noise Pollution on Human Being, Animal and property are as follows:

I) It decreases the efficiency of a man:- Regarding the impact of noise on human efficiency there are number of experiments which print out the fact that human efficiency increases with noise reduction. Thus human efficiency is related with noise.

II) Lack of concentration:- For better quality of work there should be concentration , Noise causes lack of concentration. In big cities, mostly all the offices are on main road. The noise of traffic or the loud speakers of different types of horns divert the attention of the people working in offices.

III) Fatigue:- Because of Noise Pollution, people cannot concentrate on their work. Thus they have to give their more time for completing the work and they feel tiring

IV) Abortion is caused: - There should be cool and calm atmosphere during the pregnancy. Unpleasant sounds make a lady of irriative nature. Sudden Noise causes abortion in females.

V) It causes Blood Pressure: - Noise Pollution causes certain diseases in human. It attacks on the person's peace of mind. The noises are recognized as major contributing factors in accelerating the already existing tensions of modern living. These tensions result in certain disease like blood pressure or mental illness etc.

VI) Temporary of permanent Deafness:- The effect of nose on audition is well recognized. Mechanics, locomotive drivers, telephone operators etc. All have their hearing. Impairment as a result of noise at the place of work. Physicists, physicians & psychologists are of the view that continued exposure to noise level above. 80 to 100 db is unsafe, loud noise causes temporary or permanent deafness.

VII) Effect on vegetation:- Poor quality of Crops:- Now is well known to all that plants are similar to human being. They are also as sensitive as man. There should be cool & peaceful environment for their better growth. Noise pollution causes poor quality of crops in a pleasant atmosphere.

VIII) Effect on animals:- Noise pollution damages the nervous system of animal. Animal looses the control of its mind. They become dangerous.

IX) Effect on property:- Loud noise is very dangerous to buildings, bridges and monuments. It creates waves which struck the walls and put the building in danger condition. It weakens the edifice of buildings.

Effect of Noise Pollution on Human Health

Noise health effects are the health consequences of elevated sound levels. Elevated workplace or other noise can cause hearing impairment, hypertension, ischemic heart disease, annoyance, sleep disturbance, and decreased school performance. Changes in the immune system and birth defects have been attributed to noise exposure, but evidence is limited. Although some presbycusis may occur naturally with age, in many developed nations the cumulative impact of noise is sufficient to impair the hearing of a large fraction of the population over the course of a lifetime. Noise exposure has also been known to induce tinnitus, hypertension, vasoconstriction and other cardiovascular impacts. Beyond these effects, elevated noise levels can create stress, increase workplace accident rates, and stimulate aggression and other anti-social behaviors. The most significant causes are vehicle and aircraft noise, prolonged exposure to loud music, and industrial noise.

Hearing loss

The mechanism of hearing loss arises from trauma to stereocilia of the cochlea, the principal fluid filled structure of the inner ear. The pinna combined with the middle ear amplifies sound pressure levels by a factor of twenty, so that extremely high sound pressure levels arrive in the cochlea, even from moderate atmospheric sound stimuli. Underlying pathology to the cochlea are reactive oxygen species, which play a significant role in noise-induced necrosis and apoptosis of the stereocilia. Exposure to high levels of noise have differing effects within a given population, and the involvement of reactive oxygen species suggests possible avenues to treat or prevent damage to hearing and related cellular structures.

The elevated sound levels cause trauma to the cochlear structure in the inner ear, which gives rise to irreversible hearing loss. A very loud sound in a particular frequency range can damage the cochlea's hair cells that respond to that range thereby reducing the ear's ability to hear those frequencies in the future. However, loud noise in any frequency range has deleterious effects across the entire range of human hearing. The outer ear (visible portion of the human ear) combined with the middle ear amplifies sound levels by a factor of 20 when sound reaches the inner ear.

Cardiovascular effects

Noise has been associated with important cardiovascular health problems. In 1999, the World Health Organization concluded that the available evidence showed suggested a weak association between long-term noise exposure above 67-70 dB(A) and hypertension. More recent studies have suggested that noise levels of 50 dB(A) at night may also increase the risk of myocardial infarction by chronically elevating cortisol production.

Fairly typical roadway noise levels are sufficient to constrict arterial blood flow and lead to elevated blood pressure; in this case, it appears that a certain fraction of the population is more susceptible to vasoconstriction. This may result because annoyance from the sound causes elevated adrenaline levels trigger a narrowing of the blood vessels (vasoconstriction), or independently through medical stress reactions. Other effects of high noise levels are increased frequency of headaches, fatigue, stomach ulcers and vertigo.

Annoyance

Because some stressful effects depend on qualities of the sound other than its absolute decibel (db) value, the annoyance associated with sound may need to be considered in regard to health effects. For example, noise from airports is typically perceived as more bothersome than noise from traffic of equal volume. Annoyance effects of noise are minimally affected by demographics, but fear of the noise source and sensitivity to noise both strongly affect the 'annoyance' of a noise. Even sound levels as low as 40 dB(A) (about as loud as a refrigerator or library) can generate noise complaints and the lower threshold for noise producing sleep disturbance is 45 dB(A) or lower. Other factors that affect the 'annoyance level' of sound include beliefs about noise prevention and the importance of the noise source, and annoyance at the cause (i.e. non-noise related factors) of the noise. Evidence regarding the impact of long-term noise versus recent changes in ongoing noise is equivocal on its impact on annoyance.

Estimates of sound annoyance typically rely on weighting filters, which consider some sound frequencies to be more important than others based on their presumed audibility to the human ear. The older dB(A) weighting filter described above is used widely in the U.S., but underestimates the impact of frequencies around 6000 Hz and at very low frequencies. The newer ITU-R 468 noise weighting filter is used more widely in Europe. The propagation of sound varies between environments; for example, low frequencies typically carry over longer distances. Therefore different filters, such as dB(B) and dB(C), may be recommended for specific situations.

When young children are exposed to speech interference levels of noise on a regular basis (the actual volume of which varies depending on distance and loudness of the speaker), there may develop speech or reading difficulties, because auditory processing functions are compromised. In particular the writing learning impairment known as dysgraphia is commonly associated with environmental stressors in the classroom.

Control of Noise Polluion : Control of Noise Pollution depends upon three factors:

1. To reduce the source of noise.
2. To put checks in path of its transmission.
3. To safeguard the receive of the noise. For this to happen vehicular traffic should be diverted away from dwelling sites. Proper designing of machines can reduce loss due to noise. Acoustic furnishing (absorbing techniques) should be extensively employed. There should be legal enforcement of restrictions on noise pollution.

Noise is measured in the unit of decibel (db) which is a tenth of the largest unit, the *Bel.* One decibel is equal to the finest sound that can be heard by the human ear. The followin are the methods to control and reduce noise pollution.

1. *At Source Control*—The gadgets are to be developed to control noise at source.
2. *Transmission Control*—The room walls can be covered with sound absorbers.
3. *To Protect Exposed Persons or Workers*—The devices as ear plugs and ear muffs can be provided to workers of industries.
4. *Plantation or Vegetation.* Trees should be planted along high ways, streets and other noise places, the big trees are good for this purpose.

Noise Pollution can negatively impact the body in significant ways, including elevated blood pressure, impaired cognitive functioning, and other effects of chronic stress. The following are effective strategies you can use to limit the negative impact of noise pollution and safeguard your health and happiness.

Limit The Noise : Your first line of defense against noise pollution is to do what you can to control your environment, and limit the noise that enters your space. The following are ways that you can limit environmental noise and blunt the effects of noise pollution:

Double-Paned Windows and Weather Stripping: If you live in a noisy city or near an airport, you can reduce noise in your home considerably by installing dual-paned windows, weather stripping, and even added insulation. As a bonus, these changes can also reduce your heating and cooling bills, and help the environment!

Reduce Workplace Noise: If you work in a noisy office, you may want to talk to your employer about taking steps to reduce office noise, which has been found to affect the health and productivity of workers.

Turn off The T.V.: When you're at home, a constant backdrop of television can have an effect on you as a distraction and potential stressor.

Make Bigger Changes: You may even consider moving or changing jobs if you experience significant levels of noise that you can't reduce in other ways. It sounds like a drastic step, but considering the toll that a noisy environment can take on your health, it may be an option to consider.

If you can't eliminate noise from your environment, you can actually create a healthier environment by replacing stress-inducing environmental sounds with more pleasing ones. For example, you can reduce the impact of airport or city noise with a white noise machine or 'sound spa'. They play sounds ranging from waterfalls to rain to babbling brooks to basic static, and these sounds mask the more environmental noises that can distract you or negatively affect your sleep. They can also make it easier to meditate or practice visualization techniques.

Additionally, you can drown out distracting sounds from a noisy office environment or neighbourhood with music from your iPod or stereo and enjoy the stress management and health benefits of music while lessening the impact of the other noise. This can also improve your mood, boost your immunity, calm your physiology, or energize you. While you're really trading some sounds for others, the sounds of nature or music can be more soothing and better for your health.

Because part of the toll of noise pollution is due to the activation of the body's stress response, it stands to reason that you can counteract some of these ill effects by regularly using techniques that can reverse your body's physiological changes that come with chronic stress. The following are some of the most effective techniques you can use:

- **Breathing Exercises**: Deep breathing and other breathing exercises work well because they can be done anywhere and are effective in calming the body's physiology in minutes. (Take a deep breath…don't you feel better already?)
- **Meditation:** Meditation is also an extremely effective stress reliever because it calms the physiology and even helps alter brain chemistry so that, over time, you are less reactive to stress as it happens.
- **Yoga** The practice of yoga is a great stress reliever because it combines breathing and meditation, and adds an element of exercise to be a stress reliever that acts on several different levels to benefit your health. It also provides a simple way to ease into meditation, for those who find it to be a bit of a challenge at first. These are the ways by which noise pollution can be prevented.

Marine Pollution

Oceans are ultimate sink of all natural and man-made pollutants. Marine ecosystems lack decomposing capacity. Artificial discharges cause localised pollution especially in coastal areas. Main sources are navigational oil discharge, grease and petroleum products, detergents, sewage and garbage including radioactive wastes.

Marine pollution occurs when harmful effects, or potentially harmful effects, can result from the entry into the ocean of chemicals, particles, industrial, agricultural and residential waste, or the spread of invasive organisms.

Most sources of marine pollution are land based. The pollution often comes from nonpoint sources such as agricultural runoff and wind blown debris.

Many potentially toxic chemicals adhere to tiny particles which are then taken up by plankton and benthos animals, most of which are either deposit or filter feeders. In this way, the toxins are concentrated upward within ocean food chains. Many particles combine chemically in a manner highly depletive of oxygen, causing estuaries to become anoxic.

When pesticides are incorporated into the marine ecosystem, they quickly become absorbed into marine food webs. Once in the food webs, these pesticides can cause mutations, as well as diseases, which can be harmful to humans as well as the entire food web.

Toxic metals can also be introduced into marine food webs. These can cause a change tissue matter, biochemistry, behaviour, reproduction, and suppress growth in marine life. Also, many animal feeds have a high fish meal or fish hydrolysate content. In this way, marine toxins can be transferred to land animals, and appear later in meat and dairy products.

Sources of Marine Pollution

Pollution is often classed as point source or nonpoint source pollution. Point source pollution occurs when there is a single, identifiable, and localized source of the pollution. An example is directly discharging sewage and industrial waste into the ocean. Pollution such as this occurs particularly in developing nations. Nonpoint source pollution occurs when the pollution comes from ill-defined and diffuse sources. These can be difficult to regulate. Agricultural runoff and wind blown debris are prime examples.

Plastic debris : Marine debris is mainly discarded human rubbish which floats on, or is suspended in the ocean. Eighty percent of marine debris is plastic - a component that has been rapidly accumulating since the end of World War II The mass of plastic in the oceans may be as high as one hundred million tonnes.

Discarded plastic bags, six pack rings and other forms of plastic waste which finish up in the ocean present dangers to wildlife and fisheries. Aquatic life can be threatened through entanglement, suffocation, and ingestion. Fishing nets, usually made of plastic, can be left or lost in the ocean by fishermen. Known as ghost nets, these entangle fish, dolphins, sea turtles, sharks, dugongs, crocodiles, seabirds, crabs, and other creatures, restricting movement, causing starvation, laceration and infection, and, in those that need to return to the surface to breathe, suffocation.

Many animals that live on or in the sea consume flotsam by mistake, as it often looks similar to their natural prey. Plastic debris, when bulky or tangled, is difficult to pass, and may become permanently lodged in the digestive tracts of these animals, blocking the passage of food and causing death through starvation or infection.

Plastics accumulate because they don't biodegrade in the way many other substances do. They will photodegrade on exposure to the sun, but they do so properly only under dry conditions, and water inhibits this process. In marine environments, photodegraded plastic disintegrates into ever smaller pieces while remaining polymers, even down to the molecular level. When floating plastic particles photodegrade down to zooplankton sizes, jellyfish attempt to consume them, and in this way the plastic enters the ocean food chain. Many of these long-lasting pieces end up in the stomachs of marine birds and animals, including sea turtles, and black-footed albatross.

Plastic debris tends to accumulate at the centre of ocean gyres. In particular, the Great Pacific Garbage Patch has a very high level of plastic particulate suspended in the upper water column. In samples taken in 1999, the mass of plastic exceeded that of zooplankton (the dominant animal life in the area) by a factor of six.

Toxic additives used in the manufacture of plastic materials can leech out into their surroundings when exposed to water. Waterborne hydrophobic pollutants collect and magnify on the surface of plastic debris, thus making plastic far more deadly in the ocean than it would be on land. Hydrophobic contaminants are also known to bioaccumulate in fatty tissues, biomagnifying up the food chain and putting pressure on apex predators. Some plastic additives are known to disrupt the endocrine system when consumed, others can suppress the immune system or decrease reproductive rates. Floating debris can also absorb persistent organic pollutants from seawater, including PCBs, DDT and PAHs. Aside from toxic effects, when ingested some of these are mistaken by the animal brain for estradiol, causing hormone disruption in the affected wildlife.

Toxins : Apart from plastics, there are particular problems with other toxins that do not disintegrate rapidly in the marine environment. Examples of persistent toxins are PCBs, DDT, pesticides, furans, dioxins and phenols. Heavy metals are metallic chemical elements that have a relatively high density and are toxic or poisonous at low concentrations. Examples are mercury, lead, nickel, arsenic and cadmium.

Such toxins can accumulate in the tissues of many species of aquatic life in a process called bioaccumulation. They are also known to accumulate in benthic environments, such as estuaries and bay muds: a geological record of human activities of the last century.

Eutrophication : Eutrophication is an increase in chemical nutrients, typically compounds containing nitrogen or phosphorus, in an ecosystem. It can result in an increase in the ecosystem's primary productivity (excessive plant growth and decay), and further effects including lack of oxygen and severe reductions in water quality, fish, and other animal populations.

The biggest culprit are rivers that empty into the ocean, and with it the many chemicals used as fertilizers in agriculture as well as waste from livestock and humans. An excess of oxygen depleting chemicals in the water can lead to hypoxia and the creation of a dead zone.

Estuaries tend to be naturally eutrophic because land-derived nutrients are concentrated where runoff enters the marine environment in a confined channel. In the ocean, there are frequent red tide algae blooms that kill fish and marine mammals and cause respiratory problems in humans and some domestic animals when the blooms reach close to shore.

In addition to land runoff, atmospheric anthropogenic fixed nitrogen can enter the open ocean. A study in 2008 found that this could account for around one third of the ocean's external (non-recycled) nitrogen supply and up to three per cent of the annual new marine biological production. It has been suggested that accumulating reactive nitrogen in the environment may have consequences as serious as putting carbon dioxide in the atmosphere.

Acidification : The oceans are normally a natural carbon sink, absorbing carbon dioxide from the atmosphere. Because the levels of atmospheric carbon dioxide are increasing, the oceans are becoming more acidic. The potential consequences of ocean acidification are not fully understood, but there are concerns that structures made of calcium carbonate may become vulnerable to dissolution, affecting corals and the ability of shellfis h to form shells.

Ffects of Marine Pollution

1) Untreated or partially treated sewage effluent, or organically rich industrial effluent such as that from fish processing plants, present a number of problems.

- Decomposition of organic matter causes a drop in dissolved oxygen, particularly in calm weather and sheltered bays. This can cause the death of marine plants and animals, and may lead to changes in biodiversity.
- Effluent, rich in nitrogen and phosphorus, results in `eutrophication' (over fertilization), which may cause algal blooms. These blooms can discolour the water, clog fish gills, or even be toxic, e.g. red tides. Microbial breakdown of dead algae can cause oxygen deficiencies.
- Pathogenic microorganisms cause gastric and ear-nose-throat infections, hepatitis, and even cholera and typhoid. Filter feeding animals (e.g. mussels, clams, oysters) concentrate pathogens in their gut, so eating shellfish from polluted waters is a health risk.
- Effects from industrial discharges in South Africa are generally limited to the area next to the discharge (the `mixing zone'). Water quality guidelines specify maximum levels of pollutants allowed in the receiving water.

2) Oil spills smother plants and animals, preventing respiration. In seabirds and mammals it can cause a breakdown in their thermal insulation. Chemical toxicity can cause behavioral changes, physiological damage, or impair reproduction. Oil pollution is an eyesore, and cleanup and subsequent disposal of oily wastes is difficult.

3) Pesticides, such as DDT, and other persistent chemicals e.g. PCBs, accumulate in the fatty tissue of animals. These chemicals can cause reproductive failure in marine mammals and birds.

4) Ships often paint their hulls with anti-fouling substances, e.g. tributyl-tin or TBT, to prevent growth of marine organisms. These substances leach into water and, in high traffic areas such as harbours and marinas, can affect animal life. There is a world wide trend towards limiting the use of TBT containing paints.

5) Plastics kill many marine animals. Turtles, for example, often swallow floating plastic bags, mistaking them for jelly- fish. Animals are often strangled when they become entangled with plastic debris.

Thermal Pollution

Thermal pollution is a rise or fall in the temperature change in a natural body of water caused by human influence. A common cause of thermal pollution is the use of water as a coolant by power plants and industrial manufacturers. When water used as a coolant is returned to the natural environment at a higher temperature the change in temperature impacts organisms by (a) decreasing oxygen supply, and (b) affecting ecosystem composition. Thermal pollution can also be caused by the release of very cold water from the base of reservoirs into warmer rivers. This affects fish (particularly eggs and larvae), macroinvertebrates and river productivity.

Sources and Methods

We can classify major sources that lead to thermal pollution to the following categories:

- power plants creating electricity from fossil fuel
- water as a cooling agent in industrial facilities
- deforestation of the shoreline
- soil erosion

Sources and Methods of Thermal Pollution

Major sources and Methods of pollution : Power plants creating electricity from fossil fuel Electricity is generated by heat stored in the fossil fuels and the stored energy creates a heat flow that drives turbine

The turbines then generate electricity All the electricity try to be organized so that it may travel in the same direction in a small line Organization of this sort is created by man and so it does not occur naturally Excess heat is then created because of the unnatural processes of creating organization

Water as cooling agent heat exchangers exchange heat with other streams in the factory because there steps which needs heat while other steps generates heat, leaving streams no options for intermediate temperature evaporate cooling steam is used in many final heating processes. Cooling by condensation generates great amount of waste heat from factory. Cooling comes from evaporation because ambient air is not saturated with water. Air discharged from cooling tower is a direct contribution to global warming

Deforestation of shoreline further contributes to the problem in two ways:

- aggravates soil erosion activity
- increases amount of light that strikes the water

Soil erosionsedimentation at lakes and streams makes the water muddy, muddy water lowers the clarity of water, with the introduction of impurities to the water, containing microbes and dissolved minerals, which increase the light absorption from the atmosphere increase light absorption will see a rise in the temperature of water from the heat energy of light.

Effects of a thermal pollution : Heat and hot water result from many industrial processes. They are in particular by-products of the activity of the power stations and nuclear. The water rejected into the marine mediums has harmful effects, primarily on the marine animal-life.

The pollution thermal has effects difficult to define with precision. One observes significant interferences of the reduction in the salinity of water and variations of this one in time and space. Several actions of such a pollution were however determined. It was in particular established that the hotter the temperature of rejected water is and more harmful is its effect on the benthic organizations.

Ecological effects of thermal pollution : Ecological effects — warm water

Warm water typically decreases the level of dissolved oxygen in the water. The decrease in levels of dissolved oxygen can harm aquatic animals such as fish, amphibians and copepods. Thermal pollution may also increase the metabolic rate of aquatic animals, as enzyme activity, resulting in these organisms consuming more food in a shorter time than if their environment were not changed. An increased metabolic rate may result in food source shortages, causing a sharp decrease in a population. Changes in the environment may also result in a migration of organisms to another, more suitable environment, and to in-imgration of fishes that normally only live in warmer waters elsewhere. This leads to competition for fewer resources; the more adapted organisms moving in may have an advantage over organisms that are not used to the warmer temperature. As a result one has the problem of compromising food chains of the old and new environments. Biodiversity can be decreased as a result.

It is known that temperature changes of even one to two degrees Celsius can cause significant changes in organism metabolism and other adverse cellular biology effects. Principal adverse changes can include rendering cell walls less permeable to necessary osmosis, coagulation of cell proteins, and alteration of enzyme metabolism. These cellular level effects can adversely affect mortality and reproduction.

Primary producers are affected by warm water because higher water temperature increases plant growth rates, resulting in a shorter lifespan and species overpopulation. This can cause an algae bloom which reduces the oxygen levels in the water. The higher plant density leads to an increased plant respiration rate because the reduced light intensity decreases photosynthesis. This is similar to the eutrophication that occurs when watercourses are polluted with leached agricultural inorganic fertilizers.

A large increase in temperature can lead to the denaturing of life supporting enzymes by breaking down hydrogen- and disulphide bonds within the quaternary structure of the enzymes. Decreased enzyme activity in aquatic organisms can cause problems such as the inability to break down lipids, which leads to malnutrition.

In limited cases, warm water has little deleterious effect and may even lead to improved function of the receiving aquatic ecosystem. This phenomenon is seen especially in seasonal waters and is known as thermal enrichment.

The added heat lowers the dissolved oxygen content and may cause serious problems for the plants and animals living there. In extreme cases, major fish kills can result. Warm water may also increase the metabolic rate of aquatic animals, as enzyme activity, meaning that these organisms will consume more food in a shorter time than if their environment was not changed. The temperature can be as high as 70 degrees Fahrenheit for freshwater, 80 degrees Fahrenheit for saltwater, and 85 degrees Fahrenheit for tropical fish.

Ecological effects – cold water : Releases of unnaturally cold water from reservoirs can dramatically change the fish and macro invertebrate fauna of rivers, and reduce river productivity. In Australia, where many rivers have warmer temperature regimes, native fish species have been eliminated, and macro invertebrate faunas have been drastically altered and impoverished. The temperatures for freshwater fish can be as low as 50 degrees Fahrenheit, saltwater 75 degrees Fahrenheit, and Tropical 80 degrees Fahrenheit.

Control on Thermal Pollution : Following are the means to reduce thermal pollution:

1. Theoretically, when efficiency of any heat engine is equal to 1.0 then it will convert 100% of heat energy to mechanical energy. So there will be no loss of heat to the environment. This is practically impossible. Rather, we should aim at maximizing the efficiency of heat engines (steam, IC, nuclear etc) so that heat loss is minimum.

2. Reduce mechanical friction in any rotating parts.

3. Avoid consuming energy more than necessity. Burn less coal, oil or gas.

4. Promote use of more nuclear energy because it will not generate Carbon di oxide.

5. One of the major cause of Global warming is increasing concentration of Carbon di oxide, leading to more green house effect. On the other hand, green plants have got the capacity to absorb Carbon di oxide. In the photo synthesis plants take water, sunlight and carbon di oxide to prpoduce their food. So, plant as many trees as possible. Massive plantation is the only solution for reducing global carb on di oxide level. Indirect effect of plantation is, It will reduce soil temperature, cause more rains, some of the carbon di oxide shall be dissolved in rain and shall go to the sea - which will ultimately form carbonate rocks and will help in the flora and fauna of the marine life.

5. If we can follow these, certainly we shall be successful in reducing thermal pollution and will be able to prevent the glaciers from melting and rising of sea levels.

Radiation is a form of energy which may be transferred from one body to other through empty space. Radiation has been of two main types. Sunlight—the form of radiation people are most familiar with-is an example of the type called electro magnetic radiation. It covers a broad spectrum of radiant energy and a wide range of physical and biological effects. The other type, called particulate radiation, consists of parts of atoms which are travailing at high speeds, often with tremendous energy.

Electromagnetic Radiation

All the different kinds of electromagnetic radiation have been nothing more than light rays of different wavelength and frequency. The fact that our eyes have been sensitive only to visible light, actually a very small part of the light spectrum, has been enough to remind us how different the world would look if, through some accident of nature, we could also see x-rays or radiowaves. The energy of electromagnetic radiation gets transmitted in packets called photons. All photons, of whatever part of the spectrum, travel at a speed of 186 282 miles per second, the speed of light.

Microwaves : Microwaves energy has been much too low to disrupt living tissues by ionization. Instead, the energy gets absorbed as oscillation energy and is converted to heat..This makes it possible to use microwaves for cooking. But the same radiations have been capable of heat injury to tissues in the body when exposed to the radiations thus making it necessary to carefully maintain the door seals of microwave ovens.

Ultraviolet Rays : Ultraviolet light is produced in abundance by the sun and can be generated artificially by the carbon arc or the mercury vapor arc. Ultraviolet radiation penetrates human skin only superficially, probably to a maximum depth of 1 millimeter, which, however small is enough to cause serious injury. It has been fortunate that the air, smoke, and dust filter out much of the ultraviolet radiation from the sun, especially the shorter wavelenths, for otherwise the damaging effects would drastically change life on earth.

X-rays : X-rays include a broad spectrum of wavelength. Therefre, the energy, penetrating power, and disruptive force of x-rays differ greatly depending on the wavelength. An "average" x-ray is having a wavelength of approximately 0.1 nanometer, a distance about equal to the diameter of a molecule of water. A sheet of paper coated with a luminescent substance (barium platinocyanide) was glowing. As the cathode tube was blocked off by cardboard, it was not at first apparent that the rays from the cathode tube could cause the luminescence.

Gamma Rays : Gamma rays are similar to x-rays, with which their wavelengths overlap, but because most gamma rays are of shorter wavelenth, they generally have higher energy and are more penetrating than x-rays, may be identical of gamma rays at the longer wavelength end to their spectrum. Gamma rays are having wavelengths of about 0.1 nanometer and shorter. Regardless of the wavelength, the term gamma rays is used only in reference to electromagnetic radiation that is emitted from the atomic nucleus of a radioactive element.

Particuate Radiation **:** Particulate radiation gets produced when one or more of the various components of the atom, for example, an electron, a proton. or a neutron, have been emitted from a radioactive element.

β-Particles **:** Beata Radiation get emitted by many of the radioactive substances during their decay to more stable states. An electron is exceedingly when an electron is rejected at high

speed from the atom of a radioactive element, it ionizes substances it collides with in its path. Thus beta particles damage tissue. But they are less penetrating than x-rays or gamma rays.

Carbon 24, phosphorus 32, hydrogen-3 (tritium), and strontium-90 are examples of beta emitters. Electrons with energy range 3 to 100 MeV can be obtained from the various types of accelerators, such as the Van-de-Graaff generator.

α-Particles : Some radioactive isotopes emit fast moving particles, called alpha particles, containing 2 protons and 2 neutrons. This is equivalent in make-up to a helium nucleus. Since alpha particles contain no electrons, they are positively charged. Alpha particles are less penetrating than either beta particles or x-rays and gamma rays. They could be stopped by a sheet of paper. Alpha radiation, therefore, is dangerous only when the emitter is in direct contact with tissue, such as when inhaled or ingested. But in such cases the alpha particles are readily absorbed by tissue, so the danger is very great.

Protons : Protons are 1,835 times heavier than electrons. Whie the energetic electron (beta particle) drives into tissue like a tiny particle of sand, the proton lumbers along like a rock, knocking off pieces of atoms and molecules at it goes. The proton does not penetrate as far as electron of the same energy, but it causes more disruptio in a small area.

Neutrons : It is a kind of nucleon, i.e. one of the building-blocks of atomic nuclei. The neutron has a mass of $1\text{-}8 \times 10^{-30}$ kg or, in the conventional units, 939.57 Mev/Cc^3, in and is about 1 fm in diameter. It is similar to the proton but is electrically neutral, hence the name, and can be understood as a compound of three particular kinds of quark.

Cosmic Rays : Cosmic rays are energetic sub-atomic particles, from the sun and outer space. They strike the earth at hight velocities, some of them penetrating several thousand feet of solid rock. Cosmic rays are part of the background radiation that affects all living organisms.

Biological Effects : The greatest danger of ionizig radiation has been that has been insidious. Specialised instruments are necessary to reveal its presence. Damaging exposure can thus take place without awareness of it at the time. As the biological damage, which may be irreversible, is done that the instant of exposure, it is too late to do much about it by the time any symptoms appear. The dangers from ionizing radiations have been of quite different nature from radiations encountered more commonly such as uv, visible radiation etc. The general biological effects of radiations will depend upon the exposure and may include the following.

Leucopenta. Continues over-exposure to penetrating radiation first get exhibited by a decrease in the total white blood cell count, called leucopenia.

Epilation. X-ray exposures can give rise to epilation and there was a time when x-ray expsures were used to remove superfluous hair, not without injury from the large dose required. In some cases hair may return after an interval of some months but with a large exposure dose, epilation has been permanent.

Sterility. X-rays can cause temporary or permanent sterility. Temporary sterility could be produced with moderate doses ad normal functions may return in a few months. Extremely large doses of penetrating radiation may cause permanent sterility

Tumour producation. Radiation is widely used in the treatment of cancer, but cancer may follow because of excessive radiation exposure.

Radiation Effects on Biological Molecules

Macromolecules may undergo structural alterations, which could be manifested as a change in physical properties affording sensitive means of detecting a reaction affecting only one or two atoms in a molecule.

The formation of chemical bonds has been another effet of radiation. In dilute solutions, it has been predominantly intramolecular, and may result in the molecule being pulled together to occupy less space, so that the viscosity of the solution gets produced. at higher concentrations of the solution gets produced. At higher concentrations of the solute, cross-linking has been mostly intermolecular between different molecules.

Breakage or degradation of a macromolecular may effect the primary structure, resulting in pieces of various sizes. Such main chain scission often takes place at one particular bond, suggesting that energy gets transferred along the chain from the site where it was absorbed to the weakest link; Disruption of the secondary and higher orders of structure may also occur, e.g., through the breakage of disulphate bridges, as in proteins, or very often by the rupture of weak hydrogen bonds.

Proteins. With proteins; direct action of radiation has been considerably more efficient than indirect; as little as 50-200 eV of energy deposited directly in a protein can inactivate in while some 50-200 hydroxyl free radials are needed indirectly, although one radical may be enough if it hits the right place. Indirect action may also produce important changes, e.g., insolubility. All enzymes could be inactivated in solution, the doses required vary widely, and some protection is afforded by sulphur comounds.

Nucleic acids. In nucleic acids, several significant changes occur, both in vitro and in vivo through in the cell there has been some protective effect from the proteins associated with the nucleci acids.

Breakage of the main strands may take place; a break in one chain only will be difficult to detect as the DNA helix is inherently rigid and will hold together. Breaks may join up to form crosslinks, within the same DNN molecule, between two DNA, or between DNA and portein. If there occurs a breakage in the chain in less than about five nucleotide units of a break in the other, a 'double break' arises and the molecule separates into two pieces because there are now too few hydrogen bonds between the breaks to hold it together.

Lipids. The chief reaction of lipids concerns the polyunsaturated fatty acids, and afects the carboin atom between the double bonds: hydrogen would be removed and a sesonating structure gets formed Peroxy 1 radicals and organic peroxides are radily produced in the presence of oxygen, often by means of free-radical chain reactions affecting form 20 to 1000 molecules.

Carbohydrates. Radiochemical studies of carbohydrates have been few and inconclusive. Irradiation of sugars in dilute solutions. in the absence of oxygen, causes polyment producation. This may possibly be due to cross-linking. Extremely high doses can produce degradation also.

DNA complexed with proteins has been considerably less sensitive to base destruction or chain breakage by radiation than has been non-complexed DNA. Part of this 'protecton' is from competition for radicals, but a large 'protection' is from competition for radicals, but a large fraction is also due to the structure of the DNA protein complex. Protein gets wrapped around the DNA giving a shielding effect.

Interference with DNA is almost certainly the most important cause of long-term and genetic radiation effects. Radiation effects might be able to follow the interruption of energys upplies: blocking of key enzymes might stop normal metabolism, so that the body would have to use other energy sources which might be quickly exhausted.

Radiation might break down intracellular barriers. This the so called membrane damage or enzyme release hypothesis, postulated that the primary lesion caused by radiation had been the damage to the membrances around and within the cell, which could resist isolated ionizations but not the impact of everal ionizations simultaneously.

Effects of Radiations of Molecular Levels : One of the principal causes of cellular death has been the DNA damage which is induced by ionizing rays and ultraviolet light. The study or radiation induced damages in DNA thus becomes important. The effects could be appreciated in terms of the DNA structure of the Waston and Crick. The most important feature of the structure of DNA molecule has been that it is having two polynucleotide chains which are would up together to form a double helix and held together by specific bonds between purines and pyrimidines in different chains.

The general features of DNA structure depend mainly on two main types of chemical bonding viz., the covalent polyester bonds between the nucleotides wich form a single polynucleotied chain and the hydrogen bonds between the purines and pyrimidine ring systems of two different polynucleotide chain which produce and stabilize the double helical structure. The effects produced on DNA are dependent on the nature of radiation, ionizing or ultraviolet.

Radioactive pollution : Radioactive pollution can be defined as the release of radioactive substances or high-energy particles into the air, water, or earth as a result of human activity, either by accident or by design. The sources of such waste include: 1) nuclear weapon testing or detonation; 2) the nuclear fuel cycle, including the mining, separation, and production of nuclear materials for use in nuclear power plants or nuclear bombs; 3) accidental release of radioactive material from nuclear power plants. Sometimes natural sources of radioactivity, such as radon gas emitted from beneath the ground, are considered pollutants when they become a threat to human health.

Since even a small amount of radiation exposure can have serious (and cumulative) biological consequences, and since many radioactive wastes remain toxic for centuries, radioactive pollution is a serious environmental concern even though natural sources of radioactivity far exceed artificial ones at present.

The problem of radioactive pollution is compounded by the difficulty in assessing its effects. Radioactive waste may spread over a broad area quite rapidly and irregularly (from an abandoned dump into an aquifer, for example), and may not fully show its effects upon humans and organisms for decades in the form of cancer or other chronic diseases.

Solid Waste Management

Waste management is the collection, transport, processing, recycling or disposal of waste materials. The term usually relates to materials produced by human activity, and is generally undertaken to reduce their effect on health, the environment or aesthetics. Waste management is also carried out to recover resources from it. Waste management can involve solid, liquid, gaseous or radioactive substances, with different methods and fields of expertise for each.

Waste management practices differ for developed and developing nations, for urban and rural areas, and for residential and industrial, producers. Management for non-hazardous residential and institutional waste in metropolitan areas is usually the responsibility of local government authorities, while management for non-hazardous commercial and industrial waste is usually the responsibility of the generator.

Waste management methods vary widely between areas for many reasons, including type of waste material, nearby land uses, and the area available.

Types and Sources of solid waste : Solid waste can be classified into different types depending on their source:

a) Household waste is generally classified as municipal waste,

b) Industrial waste as hazardous waste, and

c) Biomedical waste or hospital waste as infectious waste.

Municipal solid waste

Municipal solid waste consists of household waste, construction and demolition debris, sanitation residue, and waste from streets. This garbage is generated mainly from residential and commercial complexes. With rising urbanization and change in lifestyle and food habits, the amount of municipal solid waste has been increasing rapidly and its composition changing. In 1947 cities and towns in India generated an estimated 6 million tonnes of solid waste, in 1997 it was about 48 million tonnes. More than 25% of the municipal solid waste is not collected at all; 70% of the Indian cities lack adequate capacity to transport it and there are no sanitary landfills to dispose of the waste. The existing landfills are neither well equipped or well managed and are not lined properly to protect against contamination of soil and groundwater.

Garbage: the four broad categories

Organic waste: kitchen waste, vegetables, flowers, leaves, fruits.

Toxic waste: old medicines, paints, chemicals, bulbs, spray cans, fertilizer and pesticide containers, batteries, shoe polish.

Recyclable: paper, glass, metals, plastics.

Soiled: hospital waste such as cloth soiled with blood and other body fluids.

Over the last few years, the consumer market has grown rapidly leading to products being packed in cans, aluminium foils, plastics, and other such non-biodegradable items that cause incalculable harm to the environment. In India, some municipal areas have banned the use of plastics and they seem to have achieved success. For example, today one will not see a single piece of plastic in the entire district of Ladakh where the local authorities imposed a ban on plastics in 1998. Other states should follow the example of this region and ban the use of items that cause harm to the environment. One positive note is that in many large cities, shops have begun packing items in reusable or biodegradable bags. Certain biodegradable items can also be composted and reused. In fact proper handling of the biodegradable waste will considerably lessen the burden of solid waste that each city has to tackle.

There are different categories of waste generated, each take their own time to degenerate (as illustrated in the table below).

The type of litter we generate and the approximate time it takes to degenerate

Type of litter	Approximate time it takes to degenerate the litter
	Organic waste such as vegetable and fruit peels, leftover foodstuff, etc. a week or two.
Paper	10–30 days
Cotton cloth	2–5 months
Wood	10–15 years
Woolen items	1 year
	Tin, aluminium, and other metal items such as cans 100–500 years
Plastic bags	one million years?
Glass bottles	undetermined

Hazardous waste

Industrial and hospital waste is considered hazardous as they may contain toxic substances. Certain types of household waste are also hazardous. Hazardous wastes could be highly toxic to humans, animals, and plants; are corrosive, highly inflammable, or explosive; and react when exposed to certain things e.g. gases. India generates around 7 million tonnes of hazardous wastes every year, most of which is concentrated in four states: Andhra Pradesh, Bihar, Uttar Pradesh, and Tamil Nadu.

Household waste that can be categorized as hazardous waste include old batteries, shoe polish, paint tins, old medicines, and medicine bottles.

Hospital waste contaminated by chemicals used in hospitals is considered hazardous. These chemicals include formaldehyde and phenols, which are used as disinfectants, and mercury, which is used in thermometers or equipment that measure blood pressure. Most hospitals in India do not have proper disposal facilities for these hazardous wastes.

In the industrial sector, the major generators of hazardous waste are the metal, chemical, paper, pesticide, dye, refining, and rubber goods industries.Direct exposure to chemicals in hazardous waste such as mercury and cyanide can be fatal.

Hospital waste

Hospital waste is generated during the diagnosis, treatment, or immunization of human beings or animals or in research activities in these fields or in the production or testing of biologicals. It may include wastes like sharps, soiled waste, disposables, anatomical waste, cultures, discarded medicines, chemical wastes, etc. These are in the form of disposable syringes, swabs, bandages, body fluids, human excreta, etc. This waste is highly infectious and can be a serious threat to human health if not managed in a scientific and discriminate manner. It has been roughly estimated that of the 4 kg of waste generated in a hospital at least 1 kg would be infected.

Surveys carried out by various agencies show that the health care establishments in India are not giving due attention to their waste management. After the notification of the Bio-medical Waste (Handling and Management) Rules, 1998, these establishments are slowly streamlining the process of waste segregation, collection, treatment, and disposal. Many of the larger hospitals have either installed the treatment facilities or are in the process of doing so.

Prevention of solid waste

If we can reduce the amount of waste we produce there will be less adverse effects from waste generation and disposal.

City and district councils have day-to-day responsibility for waste collection and management, including waste minimisation. They prepare waste management plans following the waste management hierarchy of:

- Reduce - support products that will produce less waste, for example, those with less packaging.
- Reuse - for example, donate unwanted clothing or household goods to opportunity shops.
- Recycle - for example, aluminium cans, paper and glass can be collected and reprocessed.
- Recovery of resources - for example kitchen and garden waste can be composted.
- Residual disposal.

Role of an Individual in Prevention of Pollution

The most important reason to promote pollution prevention is to protect our health and the environment that we all must live in. When we improve and protect the environment we are ultimately protecting ourselves, and future generations.

Unfortunately, we cannot immediately make pollutants and contaminants go away. During the manufacture of many products that we use every day, pollutants are produced. Examples from large industries include automobile products (gasoline), plastic packaging, building materials, insecticides, and medicines. Small generators also produce pollutants when they provide photo processing services, educational services, dry cleaning services and more.

There are many reasons to reduce both the amount of waste generated and the amount of toxic releases to the environment. The following will discuss five important incentives:

Economic Incentives - It pays to reduce waste.

Regulatory incentives - It's the law.

Liability incentives - It's our responsibility.s

Public benefit - It's the right thing to do.

Human Health and the Environment - It's our future.

11

SOCIAL ISSUES AND THE ENVIRONMENT

Social issues are matters which directly or indirectly affects many or all members of a society and are considered to be problems, controversies related to moral values, or both.

Social issues include poverty, violence, pollution, injustice, suppression of human rights, discrimination, and crime, as well as abortion, gay marriage, gun control, autism, and the teaching of evolution, religion, the sun to name a few.

Social issues are related to the fabric of the community, including conflicts among the interests of community members, and lie beyond the control of any one individual.

Environmental ethics is the part of environmental philosophy which considers extending the tradional boundaries of ethics from soley including humans to including the non-human world. It exerts influence on a large range of disciplines including law, sociology, theology, economics, ecology and geography.

There are many ethical decisions that human beings make with respect to the environment. For example:

1. Should we continue to clear cut forests for the sake of human consumption?
2. Should we continue to propagate?
3. Should we continue to make gasoline powered vehicles?
4. What environmental obligations do we need to keep for future generations?
5. Is it right for humans to knowingly cause the extinction of a species for the convenience of humanity?

This is a **list of environmental issues** that are due to human activity. These articles relate to the anthropogenic effects on the natural environment.

6. **Climate change** — Global warming • Fossil fuels • Sea level rise • Greenhouse gas
7. **Conservation** — Habitat destruction • Habitat fragmentation • Species extinction • Pollinator decline • Coral bleaching • Whaling • Holocene extinction event • Invasive species
8. **Dams** - Environmental impacts of dams
9. **Energy** - Energy conservation • Renewable energy • Efficient energy use • Renewable energy commercialization
10. **Genetic engineering** — Genetic pollution
11. **Intensive farming** — Overgrazing • Irrigation • Monoculture • Environmental effects of meat production
12. **Land degradation** — Land pollution • Desertification \
    Soil — Soil conservation • Soil erosion • Soil contamination • Soil salination
13. **Nanotechnology** — Nanotoxicology • Nanopollution
14. **Nuclear issues** — Nuclear fallout • Nuclear meltdown • Nuclear power
15. **Overpopulation** — Burial
16. **Ozone depletion**
17. **Pollution** — Air pollution • Light pollution • Noise pollution • Thermal pollution
    Water pollution — Acid rain • Eutrophication • Marine pollution • Ocean dumping • Oil spills • Urban runoff • Water crisis
18. **Resource depletion** — Exploitation of natural resources
    Consumerism — Consumer capitalism • Planned obsolescence

Fishing — Blast fishing • Bottom trawling • Cyanide fishing • Ghost nets • Illegal, unreported and unregulated fishing • Over fishing • Shark fining

Logging — Clear cutting • Deforestation • Illegal logging

Mining — Acid mine drainage • Mountain top removal mining

19. **Toxins** — Chlorofluorocarbons • DDT • Dioxin • Heavy metals • Herbicides • Pesticides • Toxics use reduction • Toxic waste
20. **Urban sprawl**
21. **Waste** — E-waste • Litter • Waste disposal incidents

Water Conservation

Water conservation refers to reducing the use of water. The goals of water conservation efforts include:

22. **Sustainability** - To ensure availability for future generations, the withdrawal of fresh water from an ecosystem should not exceed its natural replacement rate.
23. **Energy conservation** - Water pumping, delivery, and wastewater treatment facilities consume a significant amount of energy. In some regions of the world (for example, California).
24. **Habitat conservation** - Minimizing human water use helps to preserve fresh water habitats for local wildlife and migrating waterfowl, as well as reducing the need to build new dams and other water diversion infrastructure.

Rainwater harvesting is the gathering, or accumulating and storing, of rainwater. Traditionally, rainwater harvesting has been practiced in arid and semi-arid areas, and has provided drinking water, domestic water, water for livestock, water for small irrigation and a way to replenish ground water levels. What is rainwater harvesting? The principle of collecting and using precipitation from a catchments surface.

Artificial Recharge to Ground Water is a process by which the ground water reservoir is augmented at a rate exceeding that obtained under natural conditions or replenishment. Any man-made scheme or facility that adds water to an aquifer may be considered to be an artificial recharge system.

Rain water harvesting is essential when:

25. Surface water is inadequate to meet our demand and we have to depend on ground water.
26. Due to rapid urbanization, infiltration of rain water into the sub-soil has decreased drastically and recharging of ground water has diminished.

There are two main techniques of rain water harvesting:

27. Storage of rainwater on surface for future use.
28. Recharge to ground water.

There are many types of systems to harvest rainwater. Notable systems are systems for runoff rainwater (e.g. hillside run-off) and rooftop rainwater harvesting systems. The type used depends greatly on the purpose (domestic or industrial use) and to some extent also on economics, physical and human considerations. Generally speaking, rooftop rainwater systems are most used as they are most economical (if there is more than 254mm of precipitation a year).

Components of a Rainwater Harvesting System

A Rainwater Harvesting System comprises components of various stages - transporting rainwater through pipes or drains, filtration, and storage in tanks for reuse or recharge. The common components of a rainwater harvesting system involved in these stages are illustrated here.

(i) **Catchments**: The catchment of a water harvesting system is the surface which directly receives the rainfall and provides water to the system. It can be a paved area like a terrace or courtyard of a building, or an unpaved area like a lawn or open ground. A roof made of reinforced cement concrete (RCC), galvanised iron or corrugated sheets can also be used for water harvesting.

(ii) Coarse mesh at the roof to prevent the passage of debris

(iii) **Gutters:** Channels all around the edge of a sloping roof to collect and transport rainwater to the storage tank. Gutters can be semi-circular or rectangular and could be made using:

29. Locally available material such as plain galvanised iron sheet (20 to 22 gauge), folded to required shapes.

30. Semi-circular gutters of PVC material can be readily prepared by cutting those pipes into two equal semi-circular channels.

31. Bamboo or betel trunks cut vertically in half.

The size of the gutter should be according to the flow during the highest intensity rain. It is advisable to make them 10 to 15 per cent oversize.

Gutters need to be supported so they do not sag or fall off when loaded with water. The way in which gutters are fixed depends on the construction of the house; it is possible to fix iron or timber brackets into the walls, but for houses having wider eaves, some method of attachment to the rafters is necessary.

(iv) Conduits : Conduits are pipelines or drains that carry rainwater from the catchment or rooftop area to the harvesting system. Conduits can be of any material like polyvinyl chloride (PVC) or galvanized iron (GI), materials that are commonly available.

(v) First-flushing : A first flush device is a valve that ensures that runoff from the first spell of rain is flushed out and does not enter the system. This needs to be done since the first spell of rain carries a relatively larger amount of pollutants from the air and catchment surface.

(vi) Filter : The filter is used to remove suspended pollutants from rainwater collected over roof. A filter unit is a chamber filled with filtering media such as fibre, coarse sand and gravel layers to remove debris and dirt from water before it enters the storage tank or recharge structure. Charcoal can be added for additional filtration.

(a) Charcoal water filter : A simple charcoal filter can be made in a drum or an earthen pot. The filter is made of gravel, sand and charcoal, all of which are easily available.

(b) Sand filters : Sand filters have commonly available sand as filter media. Sand filters are easy and inexpensive to construct. These filters can be employed for treatment of water to effectively remove turbidity (suspended particles like silt and clay), colour and micro-organisms.

Watershed Management

The watershed management implies, the judicious use of all

the resources i.e. land, water, vegetation in an area for providing an answer to alleviate drought, moderate floods, prevent soil erosion, improve water availability and increase food, fodder, fuel and fiber on sustained basis. Watershed to achieve maximum production with minimum hazard to the natural resources and for the well being of people. The management should be carried out on the watershed basis. The task of watershed management includes the treatment of land by using most suitable biological and engineering measures in such a manner that, the management work must be economic and socially acceptable.

Concept of Watershed

Watershed is a geo hydrological unit or piece of land that drain at a common point. A watershed is defined as any spatial area from which rain or irrigation water is collected and drained through a common point. The watershed and drainage basin are synonymous term indicating an area surrounded by a ridge line that is drained through a single outlet.

Principles of Watershed Management

The main principles of watershed management based on resource conservation, resource generation and resource utilization are:

1. Utilizing the land based on its capability
2. Protecting fertile top soil
3. Minimizing silting up of tanks, reservoirs and lower fertile lands
4. Protecting vegetative cover throughout the year
5. In situ conservation of rain water
6. Safe diversion of gullies and construction of check dams for in creasing ground water recharge
7. In creasing cropping intensity through inter and sequence cropping.
8. Alternate land use systems for efficient use of marginal lands.
9. Water harvesting for supplemental irrigation.
10. Maximizing farm income through agricultural related activities such as dairy, poultry, sheep, and goat forming.

11. Improving infrastructural facilities for storage, transport and agricultural marketing,
12. Improving socio - economic status of farmers

Objectives of Watershed Management:

13. The term watershed management is nearly synonymous with soil and water conservation with the difference that emphasis is on flood protection and sediment control besides maximizing crop production.
14. The basic objective of watershed management is thus is thus meeting the problems of land and water use, not in terms of any one resource but on the basis that all the resources are interdependent and must, therefore, be considered together.
15. The watershed aims, ultimately, at improving standards of living of common people in the basin by increasing their earning capacity, by offering facilities such as electricity, drinking water, irrigation water, freedom from fears of floods, droughts etc.
16. The overall objectives of watershed development programmers may be outlined as:
17. Recognition of watersheds as a unit for development and efficient use of land according their land capabilities for production,
18. Flood control through small multipurpose reservoirs and other water storage structures at the head water of streams and in problem areas,
19. Adequate water supply for domestic, agricultural and industrial needs.
20. Abatement of organic, inorganic and soil pollution,
21. Efficient use of natural resources for improving agriculture and allied occupation so as to improve socio-economic conditions of the local residents, and
22. Expansion of recreation facilities such as picnic and camping sites.

Climate Change

Climate change is any long-term significant change in the "average weather" of a region or the earth as a whole. Average weather may include average temperature, precipitation and wind patterns. It involves changes in the variability or average state of the atmosphere over durations ranging from decades to millions of years. These changes can be caused by dynamic processes on Earth, external forces including variations in sunlight intensity, and more recently by hun an activities.

In recent usage, especially in the context of environmental policy, the term "climate change" usually refers to changes in modern climate. For information on temperature measurements over various periods, and the data sources available, see temperature record. For attribution of climate change over the past century, see attribution of recent climate change.

Climate changes reflect variations within the Earth's atmosphere, processes in other parts of the Earth such as oceans and ice caps, and the effects of human activity. The external factors that can shape climate are often called climate forcings and include such processes as variations in solar radiation, the Earth's orbit, and greenhouse gas concentrations.

Global warming is the increase in the average the average temperature of the Earth's near-surface air and oceans since the mid-20th century, and its projected continuation.

Climate model projections indicate that global surface temperature will likely rise a further 1.1 to 6.4 °C (2.0 to 11.5 °F) during the twenty-first century. The uncertainty in this estimate arises from use of differing estimates of future greenhouse gas emissions and from use of models with differing climate sensitivity. Another uncertainty is how warming and related changes will vary from region to region around the globe. Although most studies focus on the period up to 2100, warming is expected to continue for more than a thousand years even if greenhouse gas levels are stabilized. This results from the large heat capacity of the oceans.

Increasing global temperature will cause sea levels to rise and will change the amount and pattern of precipitation, likely including an expanse of the subtropical desert regions. Other likely effects include increases in the intensity of extreme weather events, changes in agricultural yields, modifications of trade routes, glacier retreat, species extinctions and increases in the ranges of disease vectors.

Greenhouse gases are gases in an atmosphere that absorb and emit radiation within the thermal infrared range. This process is the fundamental cause of the greenhouse effect.

In our solar system, the atmospheres of Venus, Mars and Titan also contain gases that cause greenhouse effects.

Greenhouse gases, mainly water vapour, are essential to helping determine the temperature of the Earth; without them this planet would likely be so cold as to be uninhabitable. Although many factors such as the sun and the water cycle are responsible for the Earth's weather and energy balance, if all else was held equal and stable the planet's average temperature should be considerably lower without greenhouse gases.

In order, Earth's most abundant greenhouse gases are:

water vapor
carbon dioxide
methane
nitrous oxide
ozone
CFCs

When these gases are ranked by their contribution to the greenhouse effect, the most important are:

water vapor, which contributes 36–70%
carbon dioxide, which contributes 9–26%
methane, which contributes 4–9%
ozone, which contributes 3–7%

The major non-gas contributor to the Earth's green house effect, clouds, also absorb and emit infrared radiation and thus have an effect on radiative properties of the greenhouse gases.

The contribution to the greenhouse effect by a gas is affected by both the characteristics of the gas and its abundance. For example, on a molecule-for-molecule basis methane is a much stronger greenhouse gas than carbon dioxide, but it is present in much smaller concentrations so that its total contribution is smaller.

It is not possible to state that a certain gas causes an exact percentage of the greenhouse effect, because the influences of the various gases are not additive. The higher ends of the ranges quoted are for the gas alone; the lower ends, for the gas counting overlaps. Other greenhouse gases include sulfur hexafluoride, hydrofluorocarbons and perfluorocarbons. Some greenhouse gases are not often listed. For example, nitrogen trifluoride has a high global warming potential (GWP) but is only present in very small quantities.

Although contributing to many other physical and chemical reactions, the major atmospheric constituents, nitrogen (N_2), oxygen (O_2), and argon (Ar), are not greenhouse gases. This is because homonuclear diatomic molecules such as N_2 and O_2 and monatomic molecules such as Ar have no net change in their dipole moment when they vibrate and hence are almost totally unaffected by infrared light. Although heteronuclear diatomics such as carbon monoxide (CO) or hydrogen chloride (HCl) absorb IR, these molecules are short lived in the atmosphere owing to their reactivity and solubility. As a consequence they do not contribute significantly to the greenhouse effect and are not often included when discussing greenhouse gases.

Late 19th century scientists experimentally discovered that N_2 and O_2 did not absorb infrared radiation (called, at that time, "dark radiation") and that CO_2 and many other gases did absorb such radiation. It was recognized in the early 20th century that the greenhouse gases in the atmosphere caused the Earth's overall temperature to be higher than it would be without them.

Acid Rain

Acid rain is rain or any other form of precipitation that is unusually acidic. It has harmful effects on plants, aquatic animals, and infrastructure. Acid rain is mostly caused by human emissions of sulfur and nitrogen compounds which react in the atmosphere to produce acids. In recent years, many governments have introduced laws to reduce these emissions.

"Acid rain" is a popular term referring to the deposition of wet (rain, snow, sleet, fog and cloud water, dew) and dry (acidifying particles and gases) acidic components. A more accurate term is "acid deposition". Distilled water, which contains no carbon dioxide, has a neutral pH of 7. Liquids with a pH less than 7 are acidic, and those with a pH greater than 7 are basic. "Clean" or unpolluted rain has a slightly acidic pH of about 5.2, because carbon dioxide and water in the air react together to form carbonic acid, a weak acid (pH 5.6 in distilled water), but unpolluted rain also contains other chemicals.

$$H_2O\ (l) + CO_2\ (g)\ (\rightarrow)\ H_2CO_3\ (aq)$$

Carbonic acid then can ionize in water forming low concentrations of hydronium ions:

$$2H_2O\ (l) + H_2CO_3\ (aq)\ (\rightarrow)\ CO_3^{2-}\ (aq) + 2H_3O^+(aq)$$

The extra acidity in rain comes from the reaction of primary air pollutants, primarily sulfur oxides and nitrogen oxides, with water in the air to form strong acids (like sulfuric and nitric acid). The main sources of these pollutants are industrial power-generating plants and vehicles.

Natural Phenomenon of Formation of Acid Rain

The principal natural phenomena that contribute acid-producing gases to the atmosphere are emissions from volcanoes and those from biological processes that occur on the land, in wetlands, and in the oceans. The major biological source of sulfur containing compounds is dimethyl sulfide.

Acidic deposits have been detected in glacial ice thousands of years old in remote parts of the globe.

Human activity

The principal cause of acid rain is sulfur and nitrogen compounds from human sources, such as electricity generation, factories, and motor vehicles. Coal power plants are one of the most polluting. The gases can be carried hundreds of kilometers in the atmosphere before they are converted to acids and deposited. In the past, factories had short funnels to let out smoke, but this caused many problems locally; thus, factories now have taller smoke funnels. However, dispersal from these taller stacks causes pollutants to be carried farther, causing widespread ecological damage.

Chemical processes

Gas phase chemistry : In the gas phase sulfur dioxide is oxidized by reaction with the hydroxyl radical via an intermolecular reaction:

$SO_2 + OH\cdot (\rightarrow) HOSO_2\cdot$

which is followed by:

$HOSO_2\cdot + O_2 (\rightarrow) HO_2\cdot + SO_3$

In the presence of water, sulfur trioxide (SO_3) is converted rapidly to sulfuric acid:

$SO_3(g) + H_2O(l) (\rightarrow) H_2SO_4(l)$

Nitric acid is formed by the reaction of OH with nitrogen dioxide:

$NO_2 + OH\cdot (\rightarrow) HNO_3$

Chemistry in cloud droplets : When clouds are present the loss rate of SO_2 is faster than can be explained by gas phase chemistry alone. This is due to reactions in the liquid water droplets

Hydrolysis

Sulfur dioxide dissolves in water and then, like carbon dioxide, hydrolyses in a series of equilibrium reactions:

$SO_2\ (g) + H_2O \rightarrow SO_2{\cdot}H_2O$

$SO_2{\cdot}H_2O \rightarrow H^+ + HSO_3^-$

$HSO_3^- \rightarrow H^+ + SO_3^{2-}$

Oxidation

There are a large number of aqueous reactions that oxidize sulfur from S (IV) to S (VI), leading to the formation of sulfuric acid. The most important oxidation reactions are with ozone, hydrogen peroxide and oxygen (reactions with oxygen are catalyzed by iron and manganese in the cloud droplets).

Effects of Acid Rain

Acid rain has been shown to have adverse impacts on forests, freshwaters and soils, killing insect and aquatic life-forms as well as causing damage to buildings and having impacts on human health.

Surface waters and aquatic animals

Both the lower pH and higher aluminum concentrations in surface water that occur as a result of acid rain can cause damage to fish and other aquatic animals. At pHs lower than 5 most fish eggs will not hatch and lower pHs can kill adult fish. As lakes and rivers become more acidic biodiversity is reduced. Acid rain has eliminated insect life and some fish species, including the lakes, streams, and creeks in geographically sensitive areas, However, the extent to which acid rain contributes directly or indirectly via runoff from the catchment to lake and river acidity (i.e., depending on characteristics of the surrounding watershed) is variable.

Soils

Soil biology and chemistry can be seriously damaged by acid rain. Some microbes are unable to tolerate changes to low pHs and are killed. The enzymes of these microbes are denatured (changed in shape so they no longer function) by the acid. The hydronium ions of acid rain also mobilize toxins, e.g. aluminium, and leach away essential nutrients and minerals.

$$2H^+ (aq) + Mg^{2+} (clay) \rightarrow 2H^+ (clay) + Mg^{2+} (aq)$$

Soil chemistry can be dramatically changed when base cations, such as calcium and magnesium, are leached by acid rain thereby affecting sensitive species, such as sugar maple

Forests and other vegetation

Acid rain can also cause damage to certain building materials and historical monuments. This results when the sulfuric acid in the rain chemically reacts with the calcium compounds in the stones (limestone, sandstone, marble and granite) to create gypsum, which then flakes off.

$$CaCO_3 (s) + H_2SO_4 (aq) \rightarrow CaSO_4 (aq) + CO_2 (g) + H_2O (l)$$

This result is also commonly seen on old gravestones where the acid rain can cause the inscription to become completely illegible. Acid rain also causes an increased rate of oxidation for iron.

Visibility is also reduced by sulfate and nitrate aerosols and particles in the atmosphere.

Ozone Layer Depletion

Ozone depletion describes two distinct, but related observations: a slow, steady decline of about 4 percent per decade in the total volume of ozone in Earth's stratosphere (ozone layer) since the late 1970s, and a much larger, but seasonal, decrease in stratospheric ozone over Earth's polar regions during the same period. The latter phenomenon is commonly referred to as the **ozone hole**. In addition to this well-known stratospheric ozone depletion, there are also tropospheric ozone depletion events, which occur near the surface in polar regions during spring.

The detailed mechanism by which the polar ozone holes form is different from that for the mid-latitude thinning, but the most important process in both trends is catalytic destruction of ozone by atomic chlorine and bromine. The main source of these halogen atoms in the stratosphere is photodissociation of chlorofluorocarbon (CFC) compounds, commonly called freons, and of bromofluorocarbon compounds known as halons. These compounds are transported into the stratosphere after being emitted at the surface. Both ozone depletion mechanisms strengthened as emissions of CFCs and halons increased.

CFCs and other contributory substances are commonly referred to as **ozone-depleting substances (ODS)**. Since the ozone layer prevents most harmful UVB wavelengths (270–315 nm) of ultraviolet light (UV light) from passing through the Earth's atmosphere, observed and projected decreases in ozone have generated worldwide concern leading to adoption of the Montreal Protocol banning the production of CFCs and halons as well as related ozone depleting chemicals such as carbon tetrachloride and trichloroethane. It is suspected that a variety of biological consequences such as increases in skin cancer, damage to plants, and reduction of plankton populations in the ocean's photic zone may result from the increased UV exposure due to ozone depletion.

Ozone cycle overview

Three forms (or allotropes) of oxygen are involved in the ozone-oxygen cycle: oxygen atoms (O or atomic oxygen), oxygen gas (O_2 or diatomic oxygen), and ozone gas (O_3 or triatomic oxygen). Ozone is formed in the stratosphere when oxygen molecules photodissociate after absorbing an ultraviolet photon whose wavelength is shorter than 240 nm. This produces two oxygen atoms. The atomic oxygen then combines with O_2 to create O_3. Ozone molecules absorb UV light between 310 and 200 nm, following which ozone splits into a molecule of O_2 and an oxygen atom. The oxygen atom then joins up with an oxygen molecule to regenerate ozone. This is a continuing process which terminates when an oxygen atom "recombines" with an ozone molecule to make two O_2 molecules: $O + O_3 (\rightarrow) 2\ O_2$

The overall amount of ozone in the stratosphere is determined by a balance between photochemical production and recombination.

Ozone can be destroyed by a number of free radical catalysts, the most important of which are the hydroxyl radical (OH·), the nitric oxide radical (NO·) and atomic chlorine (Cl·) and bromine (Br·). All of these have both natural and anthropogenic (manmade) sources; at the present time, most of the OH· and NO· in the stratosphere is of natural origin, but human activity has dramatically increased the levels of chlorine and bromine. These elements are found in certain stable organic compounds, especially chlorofluorocarbons (CFCs), which may find their way to the stratosphere without being destroyed in the troposphere due to their low reactivity. Once in the stratosphere, the Cl and Br atoms are liberated from the parent compounds by the action of ultraviolet light, e.g. ('h' is Planck's constant, '½' is frequency of electromagnetic radiation)

$CFCl_3 + h½ (\rightarrow) CFCl_2 + Cl$

The Cl and Br atoms can then destroy ozone molecules through a variety of catalytic cycles. In the simplest example of such a cycle, a chlorine atom reacts with an ozone molecule, taking an oxygen atom with it (forming ClO) and leaving a normal oxygen molecule. The chlorine monoxide (i.e., the ClO) can react with a second molecule of ozone (i.e., O_3) to yield another chlorine atom and two molecules of oxygen. The chemical shorthand for these gas-phase reactions is:

$Cl + O_3 (\rightarrow) ClO + O_2$

$ClO + O_3 (\rightarrow) Cl + 2 O_2$

The overall effect is a decrease in the amount of ozone. More complicated mechanisms have been discovered that lead to ozone destruction in the lower stratosphere as well.

A single chlorine atom would keep on destroying ozone (thus a catalyst) for up to two years (the time scale for transport back down to the troposphere) were it not for reactions that remove them from this cycle by forming reservoir species such as hydrogen chloride (HCl) and chlorine nitrate ($ClONO_2$). On a per atom basis, bromine is even more efficient than chlorine at destroying ozone, but there is much less bromine in the atmosphere at present. As a result, both chlorine and bromine contribute significantly to the overall ozone depletion. Laboratory studies have shown that fluorine and iodine atoms participate in analogous catalytic cycles. However, in the Earth's stratosphere, fluorine atoms react rapidly with water and methane to form strongly-bound HF, while organic molecules which contain iodine react so rapidly in the lower atmosphere that they do not reach the stratosphere in significant quantities. Furthermore, a single chlorine atom is able to react with 100,000 ozone molecules. This fact plus the amount of chlorine released into the atmosphere by chlorofluorocarbons (CFCs) yearly demonstrates how dangerous CFCs are to the environment.

Consequences of ozone layer depletion

Since the ozone layer absorbs UVB ultraviolet light from the Sun, ozone layer depletion is expected to increase surface UVB levels, which could lead to damage, including increases in skin cancer. This was the reason for the Montreal Protocol. Although decreases in stratospheric ozone are well-tied to CFCs and there are good theoretical reasons to believe that decreases in ozone will lead to increases in surface UVB, there is no direct observational evidence linking ozone depletion to higher incidence of skin cancer in human beings. This is partly due to the fact that UVA, which has also been implicated in some forms of skin cancer, is not absorbed by ozone, and it is nearly impossible to control statistics for lifestyle changes in the populace.

Effects of ozone layer depletion on humans : UVB (the higher energy UV radiation absorbed by ozone) is generally accepted to be a contributory factor to skin cancer. In addition, increased surface UV leads to increased tropospheric ozone, which is a health risk to humans. The increased surface UV also represents an increase in the vitamin D synthetic capacity of the sunlight.

The cancer preventive effects of vitamin D represent a possible beneficial effect of ozone depletion. In terms of health costs, the possible benefits of increased UV irradiance may outweigh the burden.

1. **Basal and Squamous Cell Carcinomas** — The most common forms of skin cancer in humans, basal and squamous cell carcinomas, have been strongly linked to UVB exposure. The mechanism by which UVB induces these cancers is well understood — absorption of UVB radiation causes the pyrimidine bases in the DNA molecule to form dimers, resulting in transcription errors when the DNA replicates. These cancers are relatively mild and rarely fatal, although the treatment of squamous cell carcinoma sometimes requires extensive reconstructive surgery. By combining epidemiological data with results of animal studies, scientists have estimated that a one percent decrease in stratospheric ozone would increase the incidence of these cancers by 2%.

2. **Malignant Melanoma** — Another form of skin cancer, malignant melanoma, is much less common but far more dangerous, being lethal in about 15% - 20% of the cases diagnosed. The relationship between malignant melanoma and ultraviolet exposure is not yet well understood, but it appears that both UVB and UVA are involved. Experiments on fish suggest that 90 to 95% of malignant melanomas may be due to UVA and visible radiation whereas experiments on opossums suggest a larger role for UVB. Because

of this uncertainty, it is difficult to estimate the impact of ozone depletion on melanoma incidence. One study showed that a 10% increase in UVB radiation was associated with a 19% increase in melanomas for men and 16% for women.

3. **Increased Tropospheric Ozone** — Increased surface UV leads to increased tropospheric ozone. Ground-level ozone is generally recognized to be a health risk, as ozone is toxic due to its strong oxidant properties. At this time, ozone at ground level is produced mainly by the action of UV radiation on combustion gases from vehicle exhausts.

Effects on crops : An increase of UV radiation would be expected to affect crops. A number of economically important species of plants, such as rice, depend on cyanobacteria residing on their roots for the retention of nitrogen. Cyanobacteria are sensitive to UV light and they would be affected by its increase.

Effects on plankton : Research has shown a widespread extinction of plankton 2 million years ago that coincided with a nearby supernova. There is a difference in the orientation and motility of planktons when excess of UV rays reach earth. Researchers speculate that the extinction was caused by a significant weakening of the ozone layer at that time when the radiation from the supernova produced nitrogen oxides that catalyzed the destruction of ozone (plankton are particularly susceptible to effects of UV light, and are vitally important to marine food webs).

Adverse Effects Of Stessful Environment On Human Health : The environment in which we live can be considered as having three fundamental sets of components:

- Physical [energy of one form or another]
- Chemical [matter i.e. substances whether natural or man-made]
- Biological [living things].

Hazards can present themselves to us in various media e.g. air, water. The influence they can exert on our health is very complex and may be modulated by our genetic make up, psychological factors and by our perceptions of the risks that they present. The following deals with general environmental health hazards, and not extremes of climate, occupational hazards, hazards associated with food, most "accidents" or sexually transmitted disease. Health effects from economic and social consequences of environmental change are also not considered here.

Associations between an exposure and an adverse health effect do not, on their own, prove that the former is the cause of the latter.

Physical Hazards, and their Adverse Health Effects

Although you will have heard or read a great deal about the environmental consequences of global warming, man will probably be affected through famine, or war long before the health of the population as a whole is harmed to a serious degree by the temperature change. However increasing extremes of temperature, as a result of climatic change, could result in increased mortality even in temperate climates.

Important issues concerning physical hazards include those relating to health effects of electromagnetic radiation and ionising radiation. If one excludes the occupational environment, then noise and other physical hazards may present a nuisance to many inhabitants, and impair general well being. Environmental noise does not usually contribute to deafness but notable exceptions may include noisy discotheques and "personal stereos".

Electromagnetic radiation ranges from low frequency,relatively low energy, radiation such as radio and microwaves through to infra red, visible light, ultraviolet, X-rays and gamma rays. These last as well as other forms of radioactivity such as high energy subatomic particles (e.g. electrons - Beta rays) can cause intracellular ionisation and are therefore called ionising radiation. Exposure to ultraviolet (UV) radiation carries a increased risk of skin cancer such as melanoma, and of cataracts which are to an extent exposure related. Some pollutants such as chlorofluorocarbons (CFCs) used as refrigerants or in aerosol propellants or in the manufacture of certain plastics can damage the "ozone layer" in the higher atmosphere (stratosphere) and thus allow more UV light to reach us, and harm us directly. Ultraviolet light may also cause harm indirectly by contributing to an increase in ozone in the troposphere (the air we breathe) - see below under chemical hazards, or elsewhere in connection with air quality.

Radioactivity is associated with an exposure dependent risk of some cancers notably leukaemia. Contrary to popular belief however, most radiation to which the average person is exposed is natural in origin, and, of the man made sources, medical diagnosis and treatment is on average the largest source to the individual. A very important issue is the extent to which radon gas arising from certain rock types beneath dwellings can contribute to cancer risk. Ionising radiation from the nuclear industry and from fallout from detonations contributes less than 1% of the annual average dose to inhabitants of the U.K. The explanation for leukaemia clusters around nuclear power plants is not yet resolved. Similar clustering can occur in other parts of the country. The effect of viral infections associated with population shifts may be important but requires further study.

Non ionising electrical, magnetic or electromagnetic fields are an increasing focus of attention. The scientific evidence of adverse health effects from general environmental exposure to these fields is "not proven". If there are adverse effects yet to be proven, the risk is probably likely to be very small.

Chemical Hazards, and their Adverse Health Effects

If one includes tobacco smoke as an environmental hazard then it probably represents the single biggest known airborne chemical risk to health, whether measured in terms of death rates or ill-health (from lung cancer, other lung disease such as chronic bronchitis and emphysema, and disease of the heart, especially, and of blood vessels and other parts of the body). To a much lesser degree of risk, these adverse effects apply to non-smokers exposed passively to side stream tobacco smoke.

General airborne pollution arises from a variety of causes but can usefully be subdivided into pollution from combustion or from other sources. The image shows the silhouette of a power station - an important source of airborne products of combustion.

Combustion of coal and other solid fuels can produce smoke (containing polycyclic aromatic hydrocarbons - PAH) and sulphur dioxide besides other agents such as those also produced by:

Combustion of liquid petroleum products which can generate carbon monoxide, oxides of nitrogen and other agents. Industry and incineration can generate a wide range of products of combustion such as oxides of sulphur and nitrogen, polycyclic aromatic hydrocarbons, dioxins etc. Combustion of any fossil fuel generates varying amounts of particulate matter. It also adds to the environmental burden of carbon dioxide - an important "green house" gas but in these low concentrations it does not affect human health directly. Combustion of fuel can also generate hazardous substances in other ways, besides by chemical oxidation, such as by liberating benzene (from the "cracking" of petrol) or lead (from leaded petrol). Some of the primary pollutants such as nitrogen dioxide can, under the influence of UV light generate secondary pollutants notably ozone (an allotrope of oxygen).

Health effects of concern are asthma, bronchitis and similar lung diseases, and there is good evidence relating an increased risk of symptoms of these diseases with increasing concentration of sulphur dioxide, ozone and other pollutants. Moreover, there is increasing evidence to suggest that pollution from particulate matter at levels hitherto considered "safe" is associated with an increased risk of morbidity and mortality from cardiopulmonary disease especially in people with other risk factors (such as old age, or heart and lung disease). These concerns are the subject of a great deal of research throughout the world. Although high occupational exposures to exhaust especially from diesel, and to benzene does increase the risk of some cancers, reliable direct evidence of an increased to cancer risk to the population at large from the lower levels to which they are exposed is lacking.

Incineration can also generate hazardous substances if substances not best suited for disposal by incineration are "disposed" of in this way or if incineration is carried out at too low a temperature (for example this may generate dioxins).

Products of combustion and other harmful airborne pollutants can also arise within the home. Thus nitrogen dioxide generated by gas fires or gas cookers can contribute to an increased respiratory morbidity of those living in the houses. Certain modern building materials may liberate gases or vapours such as formaldehyde at low concentration but which might provoke mild respiratory and other symptoms in some occupants.

Water can be an important source of chemical hazards. It can leach lead from pipes especially if the water is soft. There is good epidemiological evidence that this can have a relatively small but measurable harmful effect especially on neurological function even at levels considered "acceptable". Other adverse effects can arise from chemicals added to the water.

Nitrate in water usually arising from fertiliser leaching (natural or artificial) can increase the risk of methaemoglobinaemia ('blue babies') in bottle fed infants but this is extremely rare. Although pesticides can and do leach into water, there is no evidence that the current standards for water quality are inadequate in this respect, but most standards are based on evidence other than human epidemiology which in this context is extremely difficult to conduct.

Beyond the point of supply further problems in drinking water quality may result. Thus for example water tanks containing lead may increase the burden of this metal in the water; while water softeners may increase its sodium content (can be harmful for bottle fed infants).

Deposition of solid hazardous waste can result in harmful substances leaching into water supplies, becoming airborne or being swallowed or otherwise absorbed directly (for example because of children playing on the sites). If the sites are well contained to prevent leaching into water supplies and segregated from human activity then the risk to human health is usually immeasurably small. However where the position of disposal sites and their contents are unknown and houses are proposed to be built on them or they are to be developed in other ways, extensive prior investigation may be needed in an attempt to estimate health risks.

Biological Hazards, and their Adverse Health Effects

These generally fall into two broad categories: those which produce adverse health effects through infection and those which produce adverse effects in non-infective (allergic) ways.

As regards microbiological hazards in water, substantial improvements in the health of the population have resulted historically from the supply of drinking water free from disease causing organisms such as cholera. Similar improvements can be expected in the health of the inhabitants of developing countries if microbiologically safe water is provided by avoidance of contamination, and appropriate purification including disinfection (usually by chlorination). Occasional outbreaks of waterborne infection still arise from contamination of drinking water by soiled water.

There can be other opportunities for further bacteriological contamination. Thus Legionella can grow in sumps or dead legs in the plumbing system and may then be dispersed as aerosols from showers.

Recreational water which is heavily contaminated with pathogens, notably coliform bacteria has been shown to be associated with an increased risk of gastrointestinal and other infectious illness, usually self-limiting.

So-called "clinical" waste is not merely an occupational hazard of health care workers but is becoming an increasingly more important risk, for example for children finding blood stained needles.

Many allergens such as grass pollen grains, or faecal material from house dust mites may cause attacks of asthma or "hay fever" (allergic rhinitis). There is evidence that high exposure to these allergens early in life, increases the risk of suffering from asthma later on. An increasing number of studies suggest that airborne chemical pollution can act synergistically with naturally occurring allergens and result in effects on lung function at concentrations lower than those at which either the allergen or the chemical irritant on its own would have produced an adverse effect.

Human Rights and Environment

Human Rights : **Human right refers** to the "basic rights and freedoms to which all humans are entitled." Examples of rights and freedoms which have come to be commonly thought of as human rights include civil and political rights, such as the right to life and liberty, freedom of expression, and equality before the law; and social, cultural and economic rights, including the right to participate in culture, the right to food, the right to work, and the right to education.

All human beings are born free and equal in dignity and rights. They are endowed with reason and conscience and should act towards one another in a spirit of brotherhood.

International human rights law is a system of laws, both domestic, regional and international, designed to promote human rights. Human rights law is made up of various international human rights instruments which are binding to its parties (nation-states that have ratified the treaty).

An important concept within human rights law is that of universal jurisdiction. This concept, which is not widely accepted, is that any nation is authorized to prosecute and punish violations of human rights wherever and whenever they may have occurred. Some customary peremptory norms of human rights are also recognised, and these are considered binding on all nations, even those that have not ratified the relevant treaty.

In principle human rights law is enforced on a domestic level and nation states that ratify human rights treaties commit themselves to enact domestic human rights legislations.

In addition to international human rights law, human rights law has been created on a regional level. The three regional human rights instrument that form binding human rights law to party states are: African Charter on Human and Peoples' Rights, the American Convention on Human Rights (the Americas) and the European Convention on Human Rights.

Human rights law is related to, but not the same as International Humanitarian Law and Refugee Law. War crimes, crimes against humanity and genocide have their own treaty law.

Laws

Human rights law is a system of laws, both domestic and international, designed to promote human rights.

International humanitarian law (IHL), often referred to as the **laws of war**, the laws and customs of war or the law of armed conflict, is the legal corpus "comprised of the Geneva Conventions and the Hague Conventions, as well as subsequent treaties, case law, and customary international law". It defines the conduct and responsibilities of belligerent nations, neutral nations and individuals engaged in warfare, in relation to each other and to *protected persons*, usually meaning civilians.

The **law of war** (also **law of armed conflict, LOAC**) is law concerning acceptable practices relating to war. In cases other than civil wars, it is considered an aspect of public international law (the law of nations). The laws of war are divided into two categories:

- **Jus in bello**, law concerning acceptable conduct in war.
- **Jus ad bellum**, law concerning acceptable justifications to use armed force.

The enforcement of international human rights law is the responsibility of the Nation State, and it's the primary responsibility of the State to make human rights a reality. There is currently no international court that upholds human rights law although the Council of Europe is responsible for both the European Convention on Human Rights, and the European Court of Human Rights that acts as a court of last appeal for human rights issues in member states. In practice, many human rights are very difficult to legally enforce due to the absence of consensus on the application of certain rights, the lack of relevant national legislation or of bodies empowered to take legal action to enforce them.

HUMAN RIGHTS AND ENVIRONMENT

The human rights and environmental movements share many of the same goals and concerns. Understanding of their common ground and of the knowledge, methods and resources that each group brings to its work improves the ability of both to realize their shared goal of a world that sustainably shelters and nurtures human and non-human life.

The Environment and Human Rights Resources website, containing resource information on human rights, environmental protection and the links between them, is one component of a three-year project to develop and promote the multiple connections between human rights and environmental protection, to the benefit of both.

Importance of the relationship between these two spheres

Over the years, the international community has increased its awareness on the relationship between environmental degradation and human rights abuses. It is clear that, poverty situations and human rights abuses are worsened by environmental degradation. This is for several obvious reasons;

• *Firstly,* the exhaustion of natural resources leads to unemployment and emigration to cities.

• *Secondly*, this affects the enjoyment and exercise of basic human rights. Environmental conditions contribute to a large extent, to the spread of infectious diseases. From the 4,400 million of people who live in developing countries, almost 60% lack basic health care services, almost a third of these people have no access to safe water supply.

• *Thirdly*, degradation poses new problems such as environmental refugees. Environmental refugees suffer from significant economic, socio-cultural, and political consequences. And fourthly, environmental degradation worsens existing problems suffered by developing and developed countries. Air pollution, for example, accounts for 2.7 million to 3.0 million of deaths annually and of these, 90% are from developing countries.

Environmental and human rights law have essential points in common that enable the creation of a field of cooperation between the two:

• *Firstly*, both disciplines have *deep social roots*; even though human rights law is more rooted within the collective consciousness, the accelerated process of environmental degradation is generating a new "environmental consciousness."

• *Secondly*, both disciplines have become internationalized. The international community has assumed the commitment to observe the realization of human rights and respect for the environment. From the Second World War onwards, the relationship State-individual is of pertinence to the international community. On the other hand, the phenomena brought on by environmental degradation transcends political boundaries and is of critical importance to the preservation of world peace and security. The protection of the environment is internationalized, while the State-Planet Earth relationship has become a concern of the international community.

• *Thirdly*, both areas of law tend to universalize their object of protection. Human Rights are presented as universal and the protection of the environment appears as everyone's responsibility.

Acquired Immuno Deficiency Syndrome (Aids)

Acquired immune deficiency syndrome or **acquired immunodeficiency syndrome (AIDS)** is a set of symptoms and infections resulting from the damage to the human immune system caused by the human immunodeficiency virus (HIV). This condition progressively reduces the effectiveness of the immune system and leaves individuals susceptible to opportunistic infections and tumors. HIV is transmitted through direct contact of a mucous membrane or the bloodstream with a bodily fluid containing HIV, such as blood, semen, vaginal fluid, preseminal fluid, and breast milk.

This transmission can involve anal, vaginal or oral sex, blood transfusion, contaminated hypodermic needles, exchange between mother and baby during pregnancy, childbirth, or breastfeeding, or other exposure to one of the above bodily fluids.

Symptoms of Aids : The symptoms of AIDS are primarily the result of conditions that do not normally develop in individuals with healthy immune systems. Most of these conditions are infections caused by bacteria, viruses, fungi and parasites that are normally controlled by the elements of the immune system that

HIV damages.

Opportunistic infections are common in people with AIDS. HIV affects nearly every organ system. People with AIDS also have an increased risk of developing various cancers such as Kaposi's sarcoma, cervical cancer and cancers of the immune system known as lymphomas. Additionally, people with AIDS often have systemic symptoms of infection like fevers, sweats (particularly at night), swollen glands, chills, weakness, and weight loss. The specific opportunistic infections that AIDS patients develop depend in part on the prevalence of these infections in the geographic area in which the patient lives.

Virus And Course of Infection : A **virus** (from the Latin *virus* meaning *toxin* or *poison*) is a sub-microscopic infectious agent that is unable to grow or reproduce outside a host cell. Viruses infect all cellular life. The origins of viruses are unclear: some may have evolved from plasmids—pieces of DNA that can move between cells—while others may have evolved from bacteria.

Viruses spread in many ways; different species of virus use different methods. For example, plant viruses are often transmitted from plant to plant by insects that feed on sap, such as aphids, while animal viruses can be carried by blood-sucking insects. These disease-bearing organisms are known as *vectors*. Influenza viruses are spread by coughing and sneezing, and others such as norovirus, are transmitted by the faecal-oral route, when they contaminate hands, food or water. Rotavirus is often spread by direct contact with infected children. HIV is one of several viruses that are transmitted through sex.

Not all viruses cause disease, as many viruses reproduce without causing any obvious harm to the infected organism. Some viruses such as HIV can cause life-long or chronic infections, and the viruses continue to replicate in the body despite the hosts' defence mechanisms. However, viral infections in animals usually cause an immune response, which can completely eliminate a virus. These immune responses can also be produced by vaccines that give lifelong immunity to a viral infection. Microorganisms such as bacteria also have defences against viral infection, such as restriction modification systems. Antibiotics have no effect on viruses, but antiviral drugs have been developed to treat life-threatening infections.

Controls of HIV/AIDS : The three main transmission routes of HIV are sexual contact, exposure to infected body fluids or tissues, and from mother to fetus or child during prenatal period. It is possible to find HIV in the saliva, tears, and urine of infected individuals, but there are no recorded cases of infection by these secretions, and the risk of infection is negligible.

Sexual contact : The majority of HIV infections are acquired through unprotected sexual relations between partners, one of whom has HIV. The primary mode of HIV infection worldwide is through sexual contact between members of the opposite sex.

During a sexual act, only male or female condoms can reduce the chances of infection with HIV and other STDs and the chances of becoming pregnant. The best evidence to date indicates that typical condom use reduces the risk of heterosexual HIV transmission by approximately 80% over the long-term, though the benefit is likely to be higher if condoms are used correctly on every occassion.

The male latex condom, if used correctly without oil-based lubricants, is the single most effective available technology to reduce the sexual transmission of HIV and other sexually transmitted infections. Manufacturers recommend that oil-based lubricants such as petroleum jelly, butter, and lard not be used with latex condoms, because they dissolves the latex, making the condoms porous. If necessary, manufacturers recommend using water-based lubricants.

Oil-based lubricants can however be used with polyurethane condoms.

The female condom is an alternative to the male condom and is made from polyurethane, which allows it to be used in the presence of oil-based lubricants. They are larger than male condoms and have a stiffened ring-shaped opening, and are designed to be inserted into the vagina.

The female condom contains an inner ring, which keeps the condom in place inside the vagina – inserting the female condom requires squeezing this ring. However, at present availability of female condoms is very low and the price remains prohibitive for many women.

Preliminary studies suggest that, where female condoms are available, overall protected sexual acts increase relative to unprotected sexual acts, making them an important HIV prevention strategy.

Studies on couples where one partner is infected show that with consistent condom use, HIV infection rates for the uninfected partner are below 1% per year. Prevention strategies are well-known in developed countries, but epidemiological and behavioral studies in Europe and North America suggest that a substantial minority of young people continue to engage in high-risk practices despite HIV/AIDS knowledge, underestimating their own risk of becoming infected with HIV.

Randomized controlled trials have shown that male circumcision lowers the risk of HIV infection among heterosexual men by up to 60%. It is expected that this procedure will be actively promoted in many of the countries affected by HIV, although doing so will involve confronting a number of practical, cultural and attitudinal issues.

Some experts fear that a lower perception of vulnerability among circumcised men may result in more sexual risk-taking behavior, thus negating its preventive effects. However, one randomized controlled trial indicated that adult male circumcision was not associated with increased HIV risk behavior.

Exposure to infected body fluids : Health care workers can reduce exposure to HIV by employing precautions to reduce the risk of exposure to contaminated blood. These precautions include barriers such as gloves, masks, protective eyeware or shields, and gowns or aprons which prevent exposure of the skin or mucous membranes to blood borne pathogens. Frequent and thorough washing of the skin immediately after being contaminated with blood or other bodily fluids can reduce the chance of infection. Finally, sharp objects like needles, scalpels and glass, are carefully disposed of to prevent needle stick injuries with contaminated items. Since intravenous drug use is an important factor in HIV transmission in developed countries, harm reduction strategies such as needle-exchange programmes are used in attempts to reduce the infections caused by drug abuse.

Mother-to-child transmission (MTCT) : Current recommendations state that when replacement feeding is acceptable, feasible, affordable, sustainable and safe, HIV-infected mothers should avoid breast-feeding their infant. However, if this is not the case, exclusive breast-feeding is recommended during the first months of life and discontinued as soon as possible.

Diaster Management : A **disaster** is the tragedy of a natural or human-made hazard that negatively affects society or environment.

In contemporary academia, disasters are seen as the consequence of inappropriately managed risk. These risks are the product of hazards and vulnerability. Hazards that strike in areas with low vulnerability are not considered a disaster, as is the case in uninhabited regions.

Developing countries suffer the greatest costs when a disaster hits – more than 95 percent of all deaths caused by disasters occur in developing countries, and losses due to natural disasters are 20 times greater (as a percentage of GDP) in developing countries than in industrialized countries.

A disaster can be defined as any tragic event that may involve at least one victim of circumstance, such as an accident, fire, terrorist attack, or explosion.

Types of Disaster : Generally, disasters are of two types – Natural and Manmade.

A **natural disaster** is the consequence of a natural hazard (e.g. volcanic eruption, earthquake, or landslide) which affects human activities. Human vulnerability, exacerbated by the lack of planning or appropriate emergency management, leads to financial, environmental or human losses. The resulting loss depends on the capacity of the population to support or resist the disaster, their resilience. This understanding is concentrated in the formulation:

"disasters occur when hazards meet vulnerability". A natural hazard will hence never result in a natural disaster in areas without vulnerability, e.g. strong earthquakes in uninhabited areas. The term *natural* has consequently been disputed because the events simply are not hazards or disasters without human involvement.

- Bushfires
- Cyclones
- Droughts
- Earthquakes
- Famine
- Floods
- Tsunamis
- Tornadoes
- Volcanos

Man-made disasters are events which, either intentionally or by accident cause severe threats to public health and well-being. Because their occurrence is unpredictable, man-made disasters pose an especially challenging threat that must be dealt with through vigilance, and proper preparedness and response. Information on the major sources of man-made disasters is provided here to help educate the public on their cause and effects as they relate to emergency planning.

- Bioterrorism
- Chemical Agents
- Pandemics and Diseases
- Radiation Emergencies
- Terrorism

Bioterrorism : A bioterrorism attack is the deliberate release of viruses, bacteria, or other germs (agents) used to cause illness or death in people, animals, or plants. These agents are typically found in nature, but it is possible that they could be changed to increase their ability to cause disease, make them resistant to current medicines, or to increase their ability to be spread into the environment.

Radiation Emergencies : Radiation is a form of energy that is present all around us. Different types of radiation exist, some of which have more energy than others. Amounts of radiation released into the environment are measured in units called curies. However, the dose of radiation that a person receives is measured in units called rem.

Chemical Agents : A chemical emergency occurs when a hazardous chemical has been released and the release has the potential for harming people's health. Chemical releases can be unintentional, as in the case of an industrial accident, or intentional, as in the case of a terrorist attack.

Terrorism : Terrorism is the use of force or violence against persons or property in violation of the criminal laws of the United States for purposes of intimidation, coercion, or ransom.

Pandemics and Diseases : A pandemic is a global disease outbreak. An influenza pandemic occurs when a new influenza virus emerges for which there is little or no immunity in the human population and the virus begins to cause serious illness and then spreads easily person-to-person worldwide.

Effects of Major Disasters

Disasters throughout history have had significant impact on the numbers, health status and life style of populations.

- Deaths
- Severe injuries, requiring extensive treatments
- Increased risk of communicable diseases
- Damage to the health facilities
- Damage to the water systems
- Food shortage
- Population movements

Health problems common to all Disasters

- Social reactions
- Communicable diseases
- Population displacements
- Climatic exposure
- Food and nutrition
- Water supply and sanitation

- Mental health
- Damage to health infrastructure

Types of Natural Disaster

Earthquake : Earthquake is an unexpected and rapid shaking of earth due to the breakage and shifting of underneath layers of Earth. Earthquake strikes all of a sudden at any time of day or night and quite violently. It gives no prior warning. If it happens in a populated area, the earthquake can cause great loss to human life and property.

Tornado : Tornado is one of the most violent storms on earth. It seems like a rotating and funnel shape cloud. It expands from the thunderstorm to the ground in the form of whirl winds reaching around 300 miles per hour. The damage path could move on to one mile wide and around 50 miles long. These storms can strike quickly without any warning.

Flood : Flood is also one of the most common hazards in the United States and other parts of the world. The effects of a flood can be local to a neighborhood or community. It can cast a larger impact, the whole river basin and multiple states could get affected. Every state is at its risk due to this hazard.

Water Damage : Water damage has a huge effect on your home, its neighborhood and your city. It is very much necessary that you should prepare for water damage. You must know what should be done during and after water damage.

Hail : Hail comes into existence when updrafts in the thunder clouds take the raindrops up towards the extremely cold regions in the atmosphere. They freeze and combine forming lumps of ice. As these lumps can be very heavy and are not supported by the updraft, they fall off with the speeds of about 100 km per hour or more. A Hail is created in the form of an enormous cloud, commonly known as thunderheads.

Wildfire : Wild forest areas catching fire is a very big problem for the people who live around these areas. The dry conditions caused several times in the year in different parts of United States can increase the possibility for wildfires. If you are well prepared in advance and know how to protect the buildings in your area, you can reduce much of the damage caused by wildfire. It is everyone's duty to protect their home and neighborhood from wildfire.

Hurricane : Hurricane also like the tornado is a wind storm, but it is a tropical cyclone. This is caused by a low pressure system that usually builds in the tropical. Huricanes comes with thunderstorms and a counterclockwise spread of winds near the surface of the earth.

Winter Freeze : Winter freeze storms are serious threats for people and their property. They include, snow, frozen rain, strong winds and extreme cold. Many precautions have to be taken in order to protect yourself, your family, home or property.

Lightning : Lightning is a much underestimated killer. Lightning is an abrupt electric expulsion which comes from cloud to cloud or from cloud to earth followed by an emission of light. Lightning is a common phenomenon after heavy rain and can also occur around 10 miles off from rainfall. Most lightning victims are people who are captivated outdoors in summer during the afternoon and evening.

Volcano : Volcano is a mountain that has an opening downwards to the reservoir of molten rock towards the surface of earth. Volcanoes are caused by the accrual of igneous products. As the pressure caused by gases in the molted rock becomes intense, the eruption takes place. The volcanic eruption can be of two kinds, quiet or volatile. The aftermaths of a volcano include flowing lava, flat landscapes, poisonous gases and fleeing ashes and rocks.

Risk Assessment

Risk assessment is a common first step in a risk management process. Risk assessment is the determination of quantitative or qualitative value of risk related to a concrete situation and a recognized threat (also called hazard). *Quantitative risk assessment* requires calculations of two components of risk: R, the magnitude of the potential loss L, and the probability p that the loss will occur.

Risk assessment consists in an objective evaluation of risk in which assumptions and uncertainties are clearly considered and presented. Part of the difficulty of risk management is that measurement of both of the quantities in which risk assessment is concerned - potential loss and probability of occurrence - can be very difficult to measure. The chance of error in the measurement of these two concepts is large. A risk with a large potential loss and a low probability of occurring is often treated differently from one with a low potential loss and a high likelihood of occurring. In theory, both are of nearly equal priority in dealing with first, but in practice it can be very difficult to manage when faced with the scarcity of resources, especially time, in which to conduct the risk management process. Expressed mathematically,

$$R_i = L_i p(L_i) \; R_{total} = \sum_i L_i p(L_i)$$

Financial decisions, such as insurance, express loss in terms of dollar amounts. When risk assessment is used for public health or environmental decisions, loss can be quantified in a common metric,such as a country's currency, or some numerical measure of a location's quality of life. For public health and environmental decisions, loss is simply a verbal description of the outcome, such as increased cancer incidence or incidence of birth defects. In that case, the "risk" is expressed as:

$$R_i = p(L_i)$$

If the risk estimate takes into account information on the number of individuals exposed, it is termed a "population risk" and is in units of expected increased cases per a time period. If the risk estimate does not take into account the number of individuals exposed, it is termed an "individual risk" and is in units of incidence rate per a time period. Population risks are of more use for cost/ benefit analysis; individual risks are of more use for evaluating whether risks to individuals are "acceptable".

Vulnerability

Vulnerability is the susceptibility to physical or emotional injury or attack. It also means to have one's guard down, open to censure or criticism; assailable. Vulnerability refers to a person's state of being liable to succumb, as to persuasion or temptation.

A sub-category of vulnerability research is social vulnerability, where increasingly researchers are addressing some of the problems of complex human interactions, vulnerability of specific groups of people, and shocks like natural hazards, climate change, and other kinds of disruptions. The importance of the issue is indicated by the establishment of endowed chairs at university departments to examine social vulnerability.

Military Vulnerability

In military circles Vulnerability is a subset of Survivability (the others being Susceptibility and Recoverability). Vulnerability is defined in various ways depending on the nation and service arm concerned, but in general it refers to the near-instantaneous effects of a weapon attack. In some definitions Recoverability (damage control, firefighting, restoration of capability) is included in Vulnerability.

Invulnerability

Invulnerability is a common feature found in video games. It makes the player impervious to pain, damage or loss of health. It can be found in the form of "power-ups" or cheats. Generally, it does not protect the player from certain instant-death hazards, most notably "bottomless" pits from which, even if the player were to survive the fall, they would be unable to escape. As a rule, invulnerability granted by power-ups is temporary, and wears off after a set amount of time, while invulnerability cheats, once activated, remain in effect until deactivated, or the end of the level is reached. Depending on the game in question, invulnerability to damage may or may not protect the player from non-damage effects, such as being immobilized or sent flying.

Rescue Operation

Rescue refers to operations that usually involve the saving of life, or prevention of injury.

Tools used might include search dogs, search and rescue horses, helicopters, and the "Jaws of Life" and other hydraulic cutting and spreading tools used to extricate individuals from wrecked vehicles. Rescue operations are sometimes supported by special vehicles such as fire department's or EMS Heavy rescue vehicle.

Ropes and special devices can reach and remove individuals and animals from difficult locations including:

- confined space rescue
- rope rescue
- cave rescue
- fast water rescue
- ice rescue
- mines rescue
- search and rescue
- urban search and rescue
- wilderness rescue
- ski patrol
- Vehicle Rescue

Rescue operations require a high degree of training and are performed by Rescue Squads, either independent or part of larger organizations like a fire, police, military, first aid squad, or ambulance services. In The U.S., they are usually staffed by medically trained personnel as NFPA regulations require it.

Floods : A **flood** is an overflow of an expanse of water that submerges land, a deluge. In the sense of "flowing water", the word may also be applied to the inflow of the tide. Flooding may result from the volume of water within a body of water, such as a river or lake, which overflows, with the result that some of the water escapes its normal boundaries. While the size of a lake or other body of water will vary with seasonal changes in precipitation and snow melt, it is not a significant flood unless such escapes of water endangers land areas used by man like a village, city or other inhabited area.

Floods can also occur in rivers, when the strength of the river is so high it flows out of the river channel, particularly at bends or meanders and cause damage to homes and businesses along such rivers. While flood damage can be virtually eliminated by moving away from rivers and other bodies of water, since time out of mind, man has lived and worked by the water to seek sustainance and capitalize on the gains of cheap and easy travel and commerce by being near water. That humans continue to inhabit areas threatened by flood damage is only evidence that the value of being near the water far exceeds the costs of repeated periodic flooding.

Principal types of flood

Riverine floods : Flooding of a creek due to heavy monsoonal rain and high tide in Darwin, Northern Territory, Australia

- **Slow kinds:** Runoff from sustained rainfall or rapid snow melt exceeding the capacity of a river's channel. Causes include heavy rains from monsoons, hurricanes and tropical depressions, foreign winds and warm rain affecting snow pack.
- **Fast kinds:** flash flood as a result of e.g. an intense thunderstorm.

Estuarine floods :

- Commonly caused by a combination of sea tidal surges caused by storm-force winds. A storm surge, from either a tropical cyclone or an extra tropical cyclone, falls within this category.

Coastal floods : Flooding near Key West, Florida, United States from Hurricane Wilma's storm surge in October 2005

- Caused by severe sea storms, or as a result of another hazard (e.g. tsunami or hurricane). A storm surge, from either a tropical cyclone or an extra tropical cyclone, falls within this category.

Catastrophic floods :

- Caused by a significant and unexpected event e.g. dam breakage, or as a result of another hazard (e.g. earthquake or volcanic eruption).

Muddy floods :

- A muddy flood is generated by runoff on cropland.

Other : Flash flooding caused by a severe thunderstorm.

- Floods can occur if water accumulates across an impermeable surface (e.g. from rainfall) and cannot rapidly dissipate (i.e. gentle orientation or low evaporation).
- A series of storms moving over the same area.
- Dam-building beavers can flood low-lying urban and rural areas, often causing significant damage.

Typical effects

Primary effects :

- *Physical damage* - Can range anywhere from bridges, cars, buildings, sewer systems, roadways, canals and any other type of structure.
- *Casualties* - People and livestock die due to drowning. It can also lead to epidemics and diseases.

Secondary effects :

- *Water supplies* - Contamination of water. Clean drinking water becomes scarce.
- *Diseases* - Unhygienic conditions. Spread of water-borne diseases
- *Crops and food supplies* - Shortage of food crops can be caused due to loss of entire harvest. However, lowlands near rivers depend upon river silt deposited by floods in order to add nutrients to the local soil.
- *Trees* - Non-tolerant species can die from suffocation.

Tertiary/long-term effects :

- *Economic* - Economic hardship, due to: temporary decline in tourism, rebuilding costs, food shortage leading to price increase etc.

Cyclones

A **cyclone** refers to an area of closed, circular fluid motion rotating in the same direction as the Earth. This is usually characterized by inward spiraling winds that rotate counter clockwise in the Northern Hemisphere and clockwise in the Southern Hemisphere of the Earth.

Large-scale cyclonic circulations are almost always centred on areas of low atmospheric pressure. The largest low-pressure systems are cold-core polar cyclones and extratropical cyclones which lie on the synoptic scale. Warm-core cyclones such as tropical cyclones, mesocyclones, and polar lows lie within the smaller mesoscale. Subtropical cyclones are of intermediate size. Cyclones have also been seen on other planets outside of the Earth, such as Mars and Neptune.

Cyclogenesis describes the process of cyclone formation and intensification. Extra tropical cyclones form as waves in large regions of enhanced midlatitude temperature contrasts called baroclinic zones. These zones contract to form weather fronts as the cyclonic circulation closes and intensifies. Later in their life cycle, cyclones occlude as cold core systems. A cyclone's track is guided over the course of its 2 to 6 day life cycle by the steering flow of the polar or subtropical jet stream.

Weather fronts separate two masses of air of different densities and are associated with the most prominent meteorological phenomena. Air masses separated by a front may differ in temperature or humidity. Strong cold fronts typically feature narrow bands of thunderstorms and severe weather, and may on occasion be preceded by squall lines or dry lines. They form west of the circulation center and generally move from west to east. Warm fronts form east of the cyclone center and are usually preceded by stratiform precipitation and fog. They move poleward ahead of the cyclone path. Occluded fronts form late in the cyclone life cycle near the enter of the cyclone and often wrap around the storm center.

Tropical cyclogenesis describes the process of development of tropical cyclones. Tropical cyclones form due to latent heat driven by significant thunderstorm activity, and are warm core. Cyclones can transition between extratropical, subtropical, and tropical phases under the right conditions. Mesocyclones form as warm core cyclones over land, and can lead to tornado formation.

Landslides

A **landslide** (or landslip) is a geological phenomenon which includes a wide range of ground movement, such as rock falls, deep failure of slopes and shallow debris flows, which can occur in offshore, coastal and onshore environments. Although the action of gravity is the primary driving force for a landslide to occur, there are other contributing factors affecting the original slope stability. Typically, pre-conditional factors build up specific sub-surface conditions that make the area/slope prone to failure, whereas the actual landslide often requires a trigger before being released.

Causes of landslides

Landslides are caused when the stability of a slope changes from a stable to an unstable condition. A change in the stability of a slope can be caused by a number of factors, acting together or alone:.

Natural causes :

- groundwater (pore water) pressure acting to destabilize the slope
- Loss or absence of vertical vegetative structure, soil nutrients, and soil structure (e.g. after a wildfire)
- erosion of the toe of a slope by rivers or ocean waves
- weakening of a slope through saturation by snowmelt, glaciers melting, or heavy rains
- earthquakes adding loads to barely-stable slopes
- earthquake-caused liquefaction destabilizing slopes volcanic eruptions

Human causes :

- vibrations from machinery or traffic
- blasting
- earthwork which alters the shape of a slope, or which imposes new loads on an existing slope
- in shallow soils, the removal of deep-rooted vegetation that binds colluvium to bedrock

- Construction, agricultural, or forestry activities which change the amount of water which infiltrates into the soil.

Earthquakes

An **earthquake** (also known as a **tremor** or **temblor**) is the result of a sudden release of energy in the Earth's crust that creates seismic waves. Earthquakes are recorded with a seismometer, also known as a seismograph. The moment magnitude of an earthquake is conventionally reported, or the related and mostly obsolete Richter magnitude, with magnitude 3 or lower earthquakes being mostly imperceptible and magnitude 7 causing serious damage over large areas. Intensity of shaking is measured on the modified Mercalli scale.

At the Earth's surface, earthquakes manifest themselves by shaking and sometimes displacing the ground. When a large earthquake epicenter is located offshore, the seabed sometimes suffers sufficient displacement to cause a tsunami. The shaking in earthquakes can also trigger landslides and occasionally volcanic activity.

In its most generic sense, the word *earthquake* is used to describe any seismic event—whether a natural phenomenon or an event caused by humans—that generates seismic waves. Earthquakes are caused mostly by rupture of geological faults, but also by volcanic activity, landslides, mine blasts, and nuclear experiments. An earthquake's point of initial rupture is called its focus or hypocenter. The term epicenter refers to the point at ground level directly above this.

12

ENVIRONMENT AND LAWS

Environment Protection Act 1986

An Act to provide for the protection and improvement of environment and for matters connected therewith.

Whereas decisions were taken at the United Nations Conference on the Human Environment held at Stokholm in June, 1972 in which India participated, to take appropriate steps for the protection and improvement of human environment;

And whereas it is considered necessary further to implement the decisions aforesaid in so far as they relate to the protection and improvement of environment and the prevention of hazards to human beings, other living creatures, plants and property.

- This Act may be called the Environemt (Protection) Act, 1986.
- It extends to the whole of India.
- It shall come into force on such date as the Central Government may, by notification in the Official Gazette, appoint and different dates may be appointed for different provisions of this Act ad for differen areas.

(1) Subject to the provisions of this Act, Central Government shall have the power to take all such measures as it seems necessary or expedient for the purpose of protecting and improving the quality of the environement and preventing, control-over and abating environmental pollution.

(2) In particular and without prejudice to the generality of the provisions of sub-section (1), such measure may include measures with respect to all or any of the following matters, namely;

- co-ordination of actions by the State Goverments, officers and other authorities;
- Under this Act, or the rule made thereunder; or
- Under any other law for the time being in force which is relatable to the objects of this Act;
- Planning and execution of a nationwide programme for the programme for the prevention, control and abatement of environmental pollution;
- laying down standards for the quality of environment is its various aspects;
- laying down standards for emission or discharge of environmental pollutants from various sources what soever; Provided the different standards for emission or discharge may be laid down under this clause from different sources having regard to the quality or composition of the emission or discharge of environmental pollutants fromsuch sources;
- restriction of areas in which any industries, operations or processes or class of industries, operations or processes shall not carried out or processes shall not be carried out subject to certain safeguards;
- laying down procedure and safeguards for the prevention of accidents which may cause environmental pollution and remedial measures for accidents;
- laying down procedures and safeguards for the handling of hazardous substances;
- examination of such manufacturing processes, materials and substances as are likely to cause environmental pollution;
- carrying out and sponsoring investigations and research relating to problems of enviromental pollution;

- inspection of any premises, plant, equipment, machinery, manfacturing or other processes, materials or substances and giving, by order, of such directions to such authorities, officers or persons as if may consider necessary to take steps for the prevention, control and abatement of environmenta pollution;
- establishment or recognition or environmental laboratories and institutes to carry out the functions entrusted to such environmental laboratories and institutes under this Act;
- collection and dissemination of information in respect of matters relating to environmental polluation;
- preparation of manuals, codes or guides relating to the prevention, control and abatement of environmental pollution;
- such other matter as the Central Government deems necessary or expedient for the purpose of securing the effective implementation of the provisions of this Act.

The Water (Prevention and Control of Pollution) Act

The Central Board for Prevention and Control of Water Pollution (which is under the administrative control of the Department of Environment), together with the state Pollution Control Boards, completed a country-wide rapid inventory of pollution from large and medium industries. This was in execution of the "Control of Pollution at Source" programme. Minimal National Standards (MINAS) for polluting discharges from specific industries were formulated, and control measures implemented in a progressively stringent manner. About thirty percent of large and medium industries of the country have installed pollution control equipment. A network of about 120 monitoring stations to check waer pollution has been established. Zoning and classification of all the 14 major inter-state rivers have been completed to provide a basis for water quality management. The river basin-wise inventory for the Yamuna and Canga has also been completed.

(b) *Land Resource Management :* Notable efforts have been made to plan, organise and provide funds to improve desert environments. To improve ecology as also to meet the requirements of the local cattle and human population forestry has been taken up relatively extensively. Rural electrification with a view to assisting the exploitation of ground water has been promoted.

An Act to provide for the prevention and control of water pollution and the maintaining or restoring of wholesomeness of water, for the establishment, with a view to carrying out the purposes aforesaid, of Boards for the prevention and control of water pollution, for conferring on and assigning to such Board powers and functions relating there to and for matters connected therewith. Whereas it is expedite to provide for the prevention and control of water pollution and the maintaining or restoring of wholesomeness of water for the establishment, with a view to carrying out the purposes aforesaid, of Boards for the prevention and control of water pollution.

The Act may be called the Water (Prevention and Control of Pollution) Act, 1974.

It applies in the first instance to the whole of the States of Assam, Bihar, Gujarat, Haryana, Himachal Pradesh, Jammu and Kashmir, Karnataka, Kerala, Madhya Pradesh, Rajasthan, Tripura and West Bengal and the Union Territories; and it shall apply to such other State which adopts this Act by resolution passed in that behalf under clause (1) of the article 252 of the Constitution.

It shall come into force, at once in the States of Assam, Bihar, Gujarat, Haryana, Himachal Pradesh, Jammu and Kashmir, Kanataka, Kerala, madhya Pradesh, Rajasthan, Tripura and West Bengal and in the Union Territories; and in any other State which adopted this Act under clause (1) of article 252 of the Constitution on the date of such adoption and any reference in this Act to the commencement of this act shall, in relation to any State or Union Territory, mean the date on which this Act comes into force in such State of Union territory.

The Air (Prevention and Control of Pollution) Act

The Air pollution Control Act became effective from May, 1981. Programmes in progress, after the framing and notification of rules under the ACt completed, include strengthening of administrative support system, inventory of area to be declared air pollution control zones, evolution of ambient air quality standards as well as industry specific emission standard. Projects in fields such as industry, energy, irrigation and mining have been subjected to environmental impact assessments, on the basis of information provided by the project authorities and supplemented by the Department of Environment's information gathering machinery.

The Wildlife (Protection) Act

The Wildlife (Protection) Act, 1972 has been amended to prohibit trade in endangered species and derivative thereof. The National Wild life Action Plan has been implemented for a rational and modern wildlife management. A comprehensive report, "Planning and Wildlife Protected Area Network in India", had been prepared and released. The scheme "Project Tiger" has been expanded for the maintenance of a viable population of the tigers in India, with 17 Tiger" reserves in 13 states of the country covering an area of 26,643 sq.km.

The Forest Conservation Act

Forests play an important role in maintaining environmental stability and in supplying the essential requirements of the people on a renewable basis. Over they years the forests have suffered depletion due to relentless pressures arising from the increasing demand for fuelwood, fodder and timber; inadequacy of protection measures; diversion of forestlands to nonforest uses, and the tendency to look upon forests as a revenue earning resource.

Realization regarding importance of the forests has always been with us, but the first practical step came in the shape of the Forest Conservation Act, 1980 which prohibited diversion of forest lands for non-forest purposes, except with the previous permission of the Central Government. Under this Act, such permissions are generally refused. Where unavoidable for development purposes, permission is given only under condition of compensatory reforestation. This Act has been amended in 1988 to make the provisions more stringent and to enforce penalties even against officers who facilitate violation of this law. As a result of strict enforcement, the rate of diversion of forest land has been brought down to about 16,000 hectares per annum compared to the rate of 1.5 lakh hectares, per annum prior to this Act.

●●●

BIBLIOGRAPHY

1 A. Borja, M. Collins, **Oceanography and Marine Environment in the Basque Country**, Elsevier Science March 2004

2 Adkins, Leonard M., **The Appalachian Trail: A Visitor's Guide,** Birmingham, Ala.: Menasha Ridge Press, 1998

3 Adkins, Leonard M.; photographs by Joe and Monica Cook, **Wildflowers of the Appalachian Trail,** Birmingham, Ala.: Menasha Ridge Press, 1999

4 Alderman, J. Anthony, Wildflowers of the Blue Ridge Parkway, Chapel Hill: University of North Carolina Press, 1997

5 Allan Barker *(Edited by),* **Molecular Methods in Ecology**, Blackwell Publishing 2000

6 Allen, Thomas J. **The Butterflies of West Virginia and Their Caterpillars.** Pitt Series in Nature and Natural History, Pittsburgh: University of Pittsburgh Press, 1997

7 A M Mannion, **Carbon and its Domestication**, Springer 2006

8 Anand Prakash and Jagadiswari Rao, **Botanical Pesticides in Agriculture**, CRC Press November 1996

9 Anderson, Larry, **Benton MacKaye: Conservationist, Planner, and Creator of the Appalachian Trail.** Creating the North American Landscape. Baltimore: Johns Hopkins University Press, 2002

10 Anil Markandya and Kirsten Halsnaes (*Edited by*), **Climate Change and Sustainable Development: Prospects for Developing Countries**, Earthscan 2002

11 Badger, Robert L. **Geology along Skyline Drive: Shenandoah National Park, Virginia.** Helena, Mont.: Falcon Publishing. 1999

12 Barbara J. Downes, Leon A. Barmuta, Peter G. Fairweather, Daniel P. Faith, Michael J. Keough, P. S. Lake, Bruce D. Mapstone, Gerry P. Quinn, **Monitoring Ecological Impacts - Concepts and Practice in Flowing Waters**, Cambridge University Press 2002

13 Bartlett, Richard A., **Troubled Waters: Champion International and the Pigeon River Controversy.** Outdoor Tennessee Series. Knoxville: University of Tennessee Press, 1995.

14 Barnes, Thomas G. **Kentucky's Last Great Places,** Lexington: University Press of Kentucky, 2002

15 Barnes, Thomas G., and S. Wilson Francis. **Wildflowers and Ferns of Kentucky,** Lexington: University Press of Kentucky, 2004

16 Bashkin, Vladimir N. *(Edited by),* **Modern Biogeochemistry: Environmental Risk Assessment (2nd edition)**, Springer 2006

17 Berry, Wendell. **Life is a Miracle: An Essay Against Modern Superstition.** Washington, D.C.: Counterpoint, 2000

18 Bell, J. N. B.; Treshow, Michael *(Edited by)*, **Air Pollution and Plant Life - 2nd Edition**, John Wiley & Sons, May 2002

19 Beniston, M., **Climatic Change and its Impacts - An overview focusing on Switzerland**, Springer 2004

20 Bingham, Nick;Blowers, Andrew;Belshaw, Chris (*Edited by*), **Contested Environments**, John Wiley & Sons May 2003

21 Bob Buchanan, Wilhelm Gruissem and Russell L. Jones *(Edited by)*, **Biochemistry & Molecular Biology of Plants**, Wiley March 2002

22 Bruce Miller, **Coal Energy Systems**, Academic Press November 2004

23 Cahalan, James M., **Edward Abbey: A Life**, Tucson: University of Arizona Press, 2001

24 Calabrese, Edward J.; Kostecki, Paul T.; Dragun, James (*Edited by*), **Contaminated Soils, Sediments and Water: Science in the Real World**, Springer 2005

25 Calvin W. Rose, **An Introduction to the Environmental Physics of Soil, Water and Watersheds**, Cambridge University Press 2004

26 Camuto, Christopher, **Hunting From Home: A Year Afield in the Blue Ridge Mountains** [Va.]. New York: W.W. Norton, 2003

27 Catrinus J. Jepma and Mohan Munasinghe, **Climate Change Policy Facts, Issues and Analyses**, Cambridge University Press June 1998

28 Cen, Kefa; Chi, Yong; Wang, Fei (*Edited by*), **Challenges of Power Engineering and Environment**, Springer May 2008

29 Chew, V. Collins, **Underfoot: A Geologic Guide to the Appalachian Trail** [guidebook]. Second edition. Harpers Ferry, W.Va.: Appalachian Trail Conference, 1997

30 Clark, Jim, **West Virginia: The Allegheny Highlands** [photographs]. Englewood, Colo.: Westcliffe Publishers, 1998

31 Clark, Jim, **Mountain Memories: An Appalachian Sense of Place** [nature photography]. Foreword by Kathy Mattea. Morgantown, W.Va.: Vandalia Press, 2003

32 C. P. Summerhayes & S. A. Thorpe with forward by R.D. Ballard, **Oceanography**, Manson Publishing

33 Colin Ratledge and Bjorn Kristiansen *(Edited by)*, **Basic Biotechnology**, Cambridge University Press May 2006

34 Dann, Kevin, **"The Appalachian Trail and the American Primeval"** [trail originator Benton MacKaye's inspiration]. In Across the Great Border Fault: The Naturalist Myth in America, 35-50. New Brunswick, N.J.: Rutgers University Press, 2000

35 Davis, Donald Edward, **Where There Are Mountains: An Environmental History of the Southern Appalachians** [landmark text]. Athens: University of Georgia Press, 2000

36 Davis, Timothy, Todd A. Croteau, and Christopher H. Marston, eds, **America's National Park Roads and Parkways: Drawings from the Historic American Engineering Record,** Baltimore: Johns Hopkins University Press, 2004

37 De Hart, Allen, **The Trails of Virginia: Hiking the Old Dominion,** Third edition. Chapel Hill: University of North Carolina Press, 2003

38 Dick, David, and Eulalie C. Dick, **Rivers of Kentucky,** Foreword by Gurney Norman. North Middletown, Ky.: Plum Lick Publishing, Inc. 2001

39 Dixon, Sherie, and Bill Tanner, **"The Tallulah Gorge",** Foxfire Magazine, 1998.

40 Dodd, C. Kenneth, Jr. **The Amphibians of Great Smoky Mountains National Park,** Knoxville: University of Tennessee Press, 2004

41 Douglas, Joseph C. **"Minerals, Moonshine, and Misanthropes: The Historic Use of Caves in the Upper Cumberland"** Lexington: University Press of Kentucky, 2004

42 Duda, Mark Damian, **West Virginia Wildlife Viewing Guide,** Helena, Mont.: Falcon Publishing, 1999

43 E Kebreab, J Dijkstra, A Bannink, W J J Gerrits and J France (*Edited by*), **Nutrient Digestion and Utilization in Farm Animals: Modelling Approaches**, CABI April 2006

44 Ellis, William E., **The Kentucky River,** Lexington: University Press of Kentucky, 2000

45 Engle, Reed L, **Everything Was Wonderful: A Pictorial History of the Civilian Conservation Corps in Shenandoah National Park,** Luray, Va.: Shenandoah Natural History Association, 1999

46 Feldman, David Lewis, and Lyndsay Moseley, **Faith-Based Environmental Initiatives in Appalachia: Connecting Faith, Environmental Concern and Reform,** World Views: Environment, Culture, Religion 7 (November), 2003

47 Feldman, David Lewis, **Going Thirsty? Appalachia Faces Water Supply Problems,** Now and Then: The Appalachian Magazine 18 (Spring), 2001

48 Ford, W. Mark, Michael A. Menzel, and Richard H. Odom, **Elevation, Aspect, and Cove Size Effects on Southern Appalachian Salamanders,** Southeastern Naturalist 1 (December), 2002

49 Fortney, Ronald H., James Rentch, Harold S. Adams, and Steven Stephenson, **Vegetation of the Bluestone River Gorge in Southern West Virginia,** In Proceedings, New River Symposium, April 11-12, 1997, Glade Springs Resort, Daniels,

West Virginia, 11-19. Glen Jean, W.Va.: National Park Service, 1997

50 Fraser, Rory, Steve Hollenhorst, and Alan Collins, **Shades of Green: Public Opinion and the Economics of Environmental Protection in West Virginia,** In Inside West Virginia: Public Policy Perspectives for the 21 st Century, eds. B. Keith and R. Althouse, 107-133. Morgantown: West Virginia University Press,1999

51 Frome, Michael, **Protecting the Public Options,** In Greenspeak: Fifty Years of Environmental Muckraking and Advocacy, by M. Frome, 66-80. Knoxville: University of Tennessee Press, 2002

52 Gaddy, L. L. **A Naturalist's Guide to the Southern Blue Ridge Front: Linville Gorge, North Carolina, to Tallulah Gorge, Georgia,** Columbia: University of South Carolina Press, 2000

53 Garland, Mark S., and John Anderton, illustrator, **Watching Nature: A Mid-Atlantic Natural History,** Washington, D.C.: Smithsonian Institution Press, 1997

54 Garnette, Bill, **A Spruce Knob Miracle,** Goldenseal: West Virginia Traditional Life 26 (Fall), 2000

55 Gilfillan, Merrill, **Burnt House to Paw Paw: Appalachian Notes,** Profile Series, 3. Stockbridge, Mass.: Hard Press, 1997

56 Hawksworth, David L.; Bull, Alan T. *(Edited by),* **Biodiversity and Conservation in Europe**, Springer 2008

57 Heikki M. T. Hokkanen, James M. Lynch *(Edited by),* **Biological Control - Benefits and Risks**, Cambridge University Press March 1996

58 Helaine Selin *(Edited by),* **Nature Across Cultures: Views of Nature and the Environment in Non-Western Cultures**, Kluwer Academic Publishers September 2003

59 Helen Caldicott, **Nuclear power is not the answer to global warming or anything else,** Melbourne University Press, (2006)

60 Ian F. Spellerberg, John W. D. Sawyer, Foreword by Tony Whitten, **An Introduction to Applied Biogeography**, Cambridge University Press 1999

61 Isabelle M. Côté, John D. Reynolds (*Edited by*), **Coral Reef Conservation**, CUP August 2006

62 J Aerts, P Droogers (*Edited by*) **Climate Change in Contrasting River Basins: Adaptation Strategies for Water, Food and Environment**, CABI December 2004

63 J D van Mansvelt and M J van der Lubbe, **Checklist for Sustainable Landscape Management**, Elsevier 1999
64 John Postgate, **Nitrogen Fixation (3rd edition)**, Cambridge University Press 1998
65 Joșeph Seckbach *(Edited by)*, **Algae and Cyanobacteria in Extreme Environments**, Springer 2007
66 Kerry O. Britton *(Edited by)*, **Biological Pollution: An Emerging Global Menace**, APS Press 2004
67 L.C. Banaras, J.P. Gaur *(Edited by)*, **Algal Adaptation to Environmental Stresses - Physiological, Biochemical and Molecular Mechanisms,** Springer Verlag 2001
68 Louise E Buck, Charles C Geisler, John Schelhas, Eva Wollenberg, **Biological Diversity: Balancing Interests Through Adaptive Collaborative Management**, CRC Press 2001
69 M A K Khalil, **Atmospheric Methane - Its Role in the Global Environment**, Springer Verlag 2000
70 M A van Drunen, R Lasage and C Dorlands *(Edited by)*, **Climate Change in Developing Countries**, CABI September 2006
71 M. Niaounakis and C.P. Halvadakis, **Olive Processing Waste Management - Literature Review and Patent Survey (2nd Edition)**, Pergamon January 2006.
72 Matthew A. Tarr, **Chemical Degradation Methods for Wastes and Pollutants - Environmental and Industrial Applications**, Marcel Dekker August 2003
73 Malin Falkenmark and Johan Rockstrom*(Edited by)*, **Balancing Water for Humans and Nature - the new approach in ecohydrology**, Earthscan July 2004
74 Martin Alexander, **Biodegradation and Bioremediation (second edition)**, Academic Press 2001
75 Merian, Ernest; Anke, Manfred; Ihnat, Milan; Stoeppler, Markus, **Elements and their Compounds in the Environment - Occurrence, Analysis and Biological Relevance (in 3 Volumes)**, Wiley-VCH Jan 2004
76 Michael J Crawley *(Edited by)*, **Natural Enemies - The Population Biology of Predators, Parasites and Diseases**, Blackwell Publishing 1992

77 Morris, Dick; Freeland, Joanna; Hinchliffe, Steve; Smith, Sandy (*Edited by*), **Changing Environments**, John Wiley & Sons March 2003

78 Peter Calow, **Encyclopedia of Ecology and Environmental Management**, Blackwell Publishing 1999

79 Raluca-Ioana Stefan, Jacobus Frederick van Staden and Hassan Y Aboul-Enein, **Electrochemical Sensors in Bioanalysis**, Marcel Dekker 2001

80 Ravendra Naidu (*Edited by*), **Chemical Bioavailability in Terrestrial Environments**, Elsevier March 2008

81 Richard B Philp, **Ecosystems and Human Health: Toxicology and Environmental Hazards, Second Edition**, CRC Press 2001

82 Robert Smith, Thomas Smith, **Elements of Ecology - 6th Edition**, Pearson Education November 2005

83 Rod Barratt, **Atmospheric Dispersion Modelling: An Introduction to Practical Applications**, Earthscan 2001

84 Roger Perman, Michael Common, James Mcgilvray, Yue Ma, **Natural Resource and Environmental Economics - 3rd Edition**, Pearson March 2003

85 R Nieder and D K Benbi, **Carbon and Nitrogen in the Terrestrial Environment,** Springer July 2008

86 Sagar V. Krupa, **Air Pollution, People, and Plants**, APS 1997

87 Schweingruber, Fritz H., Börner, Annett, Schulze, Ernst-Detlef, **Atlas of Woody Plant Stems - Evolution, Structure, and Environmental Modifications**, Springer 2006

88 Seinfeld, John H.; Pandis, Spyros N., **Atmospheric Chemistry and Physics - From Air Pollution to Climate Change - 2nd edition**, John Wiley & Sons September 2006

89 Shimshon S Belkin and Rita R Colwell *(Edited by),* **Oceans and Health - Pathogens in the Marine Environment**, Springer October 2005

90 Spiros N. Agathos and Walter Reineke *(Edited by),* **Biotechnology for the Environment: Soil Remediation**, Kluwer Academic Publishers December 2002

91 Spiros N. Agathos and Walter Reineke *(Edited by)*, **Biotechnology for the Environment: Wastewater Treatment and Modeling, Waste Gas Handling**, Kluwer Academic Publishers March 2003

92 Stephen J Culver, **Biotic Response to Global Change**, Cambridge University Press July 2000

93 Steffen, W., Jäger, J.,Carson, D. J., Bradshaw, C. (*Edited by*), **Challenges of a Changing Earth**, Springer 2003

94 T Addiscott *(Edited by)*, **Nitrate, Agriculture and the Environment**, CABI July 2005

95 Thad Godish, **Air Quality, Fourth Edition**, CRC Press 2003

96 Tom Fenchel, Gary King and George, **Bacterial Biogeochemistry - The Ecophysiology of Mineral Cycling (second edition)**, Academic Press 1998

97 Tor Bjorn Larsson *(Edited by)*, **Biodiversity Evaluation Tools for European Forests**, Blackwell Publishing 2001

98 Van Oppen, Madeleine J. H.; Lough, Janice M. (*Edited by*), **Coral Bleaching**, Springer January 2009

99 Vincent Hegarty, **Nutrition, Food and the Environment**, AACC Press 1995

100 W Barthlott and M Winiger *(Edited by)*, **Biodiversity - A Challenge for Development Research and Policy**, Springer 2001

101 Werner, Dietrich; Newton, William E. (*Edited by*), **Nitrogen Fixation in Agriculture, Forestry, Ecology, and the Environment**, Springer 2005

102 W Nentwig *(Edited by)*, **Biological Invasions**, Springer January 2007

103 William N Nierenberg (*Edited by*), **Encyclopedia of Environmental Biology**, Academic Press 1995

W. Niessen (*Edited by*), **Current Practice of Gas Chromatography-Mass Spectrometry**, Marcel Dekker 2001

●●●